NEW TOOLS, OLD TAS

This book is dedicated to my loved ones:
Linda, Maja Anneli, Torkel, mother, father, Hanne and Grete.

New Tools, Old Tasks
Safety Implications of New Technologies and Work Processes for Integrated Operations in the Petroleum Industry

TORGEIR K. HAAVIK

NTNU Social Research, Trondheim, Norway

CRC Press
Taylor & Francis Group
Boca Raton London New York

CRC Press is an imprint of the
Taylor & Francis Group, an **informa** business

CRC Press
Taylor & Francis Group
6000 Broken Sound Parkway NW, Suite 300
Boca Raton, FL 33487-2742

First issued in paperback 2017

© 2013 by Torgeir K. Haavik
CRC Press is an imprint of Taylor & Francis Group, an Informa business

No claim to original U.S. Government works

Version Date: 20160226

ISBN 13: 978-1-138-07273-2 (pbk)
ISBN 13: 978-1-4094-5029-0 (hbk)

Visit the Taylor & Francis Web site at
http://www.taylorandfrancis.com

and the CRC Press Web site at
http://www.crcpress.com

Contents

List of Figures

List of Tables

Preface

This book has been written in the intersection between different fields of interest: engineering, social sciences and the peculiarities of sociotechnical systems. Hence, the number of people with whom I have discussed the various topics of the book is considerable. Many thanks to all of them.

Petter Almklov is perhaps the single person who has influenced my thinking the most. Ever since we started our geological training more than 20 years ago, he has been a faithful provider of brilliant and idiosyncratic observations and explanations of trivial as well as exotic phenomena. If I could only see the patterns and the meaning in and behind the nonsensical as clearly as he does!

Vidar Hepsø has provided essential practical and professional support to the project from the early fieldwork to the analysing and writing process. Thanks for being an excellent discussion partner and motivator.

Studio Apertura, which has been my work place over the last five years, has been a stimulating and friendly community which has offered the mixture of professionalism and fun needed in any type of work. Many thanks to Per Morten Schiefloe, who made this study possible in the first place, and to all of my colleagues. Thanks also to NTNU Social Research for giving me the time I needed to finish the book.

The Ashgate editors have been very helpful in the whole process – thank you very much. Many thanks also to Erik Hollnagel for searching his archives to provide digital versions of several of the figures used in the book.

Last I want to thank my wife Linda for being such a committed, critical and cool reviewer of my ideas and writings.

Some of the content in this book is based on material previously published in the following articles: Haavik, T. 2010. Making Drilling Operations Visible: The Role of Articulation Work for Organisational Safety. *Cognition, Technology & Work*, 12(4), 285–295; Haavik, TK. 2011. Chasing Shared Understanding in Drilling Operations. *Cognition, Technology & Work*, 13(4), 281–294; Haavik, TK. 2011. On Components and Relations in Sociotechnical Systems. *Journal of Contingencies and Crisis Management*, 19(2), 99–109; Haavik, TK. 2012. Challenging Controversies: A Prospective Analysis of the Influence of New Technologies on the Safety of Offshore Drilling Operations. *Journal of Contingencies and Crisis Management*, 20(2), 90–101. Material from the two first articles is used with kind permission from Springer Science+Business Media B.V. Material from the two last is used with kind permission from John Wiley and Sons.

Preface

Chapter 1
Introduction

This is a book about Integrated Operations (IO), and the impact of this operating regime on the safety of the petroleum industry. It is based on several months of fieldwork in the oil and gas industry in the period 2008–2009, including observation studies, interviews and document studies, where the goal was to understand how Integrated Operations may affect operational safety.

This worldwide change process in the petroleum industry has many other names, such as Field of the future, Smart fields and iField. In this book, these are seen as different notations for similar change processes. The generic principles for Integrated Operations are familiar to a range of domains, such as the nuclear, military and aviation industry; more use of information communication technology (ICT) will enable remote control, automation and better planning processes; all in all faster, better and safer decisions and operations.

Why do we need a book on this topic? Why not simply assume that the experiences from other industries will be valid also for the petroleum industry? While the latter may be the case, it will be on a very general level. I think the greatest risk in this type of research is the risk of over-generalising, of addressing visions and management philosophies at the cost of the nitty-gritty details of the dirty reality. The consequence may be tautological arguments that are not sufficiently rooted in the empirical field they address. For example, I believe one may find many similarities between the technologies and work processes involved in the collaborative work in an air traffic management location and an onshore operations support centre for offshore drilling operations. However, the physical natures of the work domains are very different, and one cannot expect that a generic technology or work process will make the same difference to those domains.

This book is therefore devoted to exploring the implications of defined IO tools for defined work domains. Any further generalisation should be a *consequence* of such insights, not *produce* those insights. Having said that, Integrated Operations may mean different things to different people, and I am not suggesting that the modus operandi in this book is the only one possible. It is, however, one very concrete contribution to the larger work of understanding the implications of Integrated Operations for the safety of the petroleum industry.

1.1 Background and Objective

Since Norwegian oil and gas exploration and production started more than 40 years ago, the petroleum industry has achieved a central position in the Norwegian industry, and society as a whole. In economic terms the historical development is often described as an adventure. The Norwegian economy is largely founded on the oil and gas industry. The ripple effect on general working life has been extensive. A whole range of industries supply the oil and gas sector with technologies, services and knowledge, and there is a significant mutual benefit between the petroleum industry and the academic and research sector and a range of industries that are not directly connected to the oil and gas industry (see e.g. Kindingstad, 2002; Schiefloe, 2009 for a more thorough elaboration of the history of the Norwegian oil industry).

The societal dependency on the petroleum industry is matched by an equally strong dependency among consumers. Both as commodities and energy sources, oil and gas are indispensable. Norway is not a special case; 'double dependence' can be seen as a more general, characteristic feature of the international petroleum industry. It looks like a win–win situation, but the challenges associated with the petroleum industry make it a source of many and deep controversies.

The risks are well known. Drilling operations involve intervention in pressurised petroleum reservoirs, and as history tells us, blowouts represent a severe threat to lives, environment and materials. This book addresses the challenges associated with drilling operations. It does so, however, by addressing the *construction of safe operations rather than the prevention of accidents*. The well construction process implies drilling through a series of geological zones, some of which may contain gases and fluids under high pressure. Safe operation in such a hostile subsurface environment requires continuous interpretation work to comprehend the subsurface conditions, deploy adequate drilling strategies, and to avoid the many drilling risks amongst which a blowout has the most catastrophic potential. Consequently, the notion of safety in this book refers to the work associated with the subsurface domain.

Both in Norway and internationally we have repeatedly been reminded of the risk potential of the petroleum industry: on 3 June 1979, the drilling rig Sedco 135 experienced a blowout from the exploration oil well Ixtoc I in the Gulf of Mexico, leading to one of the largest oil spills in history (Jernelöv and Lindén, 1981). The Gulf of Mexico was also the location where, on 20 April 2010, the oil rig Deepwater Horizon experienced a blowout leading to the death of eleven people and one of the worst oil catastrophes ever with respect to environmental consequences (Graham et al., 2011). The Snorre A accident that occurred on 11 November 2004 represents one of the most serious incidents ever on the Norwegian continental shelf, the unleashing of the catastrophic potential being prevented at the last minute (Brattbakk et al., 2005).[1]

In Norway, risks associated with the petroleum industry have led to strong political controversies associated with extending oil and gas exploration and

---

1   Other and more serious accidents could have been mentioned. These three are referred to because of their specific connection to the subsurface domain and thus the main empirical focus of this book.

production to areas far north where the ecosystems are particularly vulnerable. As this is being written, the controversy over petroleum exploration in the Lofoten/ Vesterålen region has still not been settled after more than a decade of struggles involving political parties, the petroleum industry and environmental organisations. Sceptics who call attention to the high risk potential are met with arguments that safety can be improved through better technology and work processes. The safety of the industry is thus a central point of controversy, and at the same time it is a point that all stakeholders agree upon; safety is a central goal for all.

Since the start of this millennium the term *Integrated Operations* has received more and more attention. In short, Integrated Operations denotes an operating regime where new technologies and work processes make possible an increased use of real-time data,[2] collaboration across disciplines and geographical distances and expert knowledge, with the goal of achieving better, faster and safer operations (see e.g. OLF, 2005). However, since the promise of safer operations needs to be sustained by more than visions, and since there still is limited practical experience with many of the more advanced IO solutions, it is a challenging research task to evaluate the effects. In this book, the research task is approached by delineating and concretising both the concept of Integrated Operations and the specific domain into which it is introduced. This strategy is crucial for the analyses to obtain a certain degree of accuracy, and to avoid too general, non-falsifiable assumptions. *The main objective book of this book is to explore how we can understand the effect of Integrated Operations on the safety of the petroleum industry in general, and on offshore drilling operations especially.*

This objective is in turn translated and operationalised into a set of four research questions that throw light on different aspects of the topic. This translation is actually an important methodical issue, since the way the research questions were formulated is rather a *result* of the fieldwork than a point of departure from which the fieldwork was designed. This relates the study to an aspect of the grounded theory tradition (Glaser, 1994; Glaser and Strauss, 1967) that allows the research questions to grow out of the fieldwork and to change as data are generated and analysed. This point is further discussed in other chapters of the book. The results of these discussions are, however, forestalled here.

The objective of the study might point in several directions with respect to the formulation of research questions as well as adoption of methodology. A large volume of existing literature and research on the topic of industrial and organisational safety, some of which is reviewed in Chapter 5, offers many adequate perspectives and discourses that are helpful to the research design. The strengths of some of these discourses, however, sometimes represent an inconsistency with the epistemological foundation[3] this study rests upon. While frameworks

---

2 Use of real-time data means that sensor data are processed (e.g. visualised or used in calculations) continuously, as they are produced.

3 Some readers may react to the use of the terms epistemology and ontology in this book. Traditionally ontology refers to *what the world actually is* and epistemology refers

and models of safety and accident genealogy may offer conceptual and intuitive representations, they also often presuppose the frameworks' population and the power relations between the actors. Sections 5.1–5.4 elaborate on some of these traditional safety/accidents/work studies frameworks, and Section 5.5 introduces an alternative approach that has so far not been greatly represented in the field of safety research.

As a result of the fieldwork and its preliminary descriptions and interpretations informed by Actor-Network-Theoretical (ANT) approaches as well as traditional safety frameworks, a set of research questions was developed. These research questions address complementary aspects of the main objective. They also constitute an argumentation where the answers to the former questions work as arguments to the next. The first question defines the field and the ontology on which the second question builds. The second and third questions point to specific consequences of such an ontology. Lastly, the fourth question draws on the findings from the first three questions and addresses the main objective of the book in a more direct, but also more generic manner.[4]

The first question is: *How should sociotechnical systems be described and understood?* This is a fundamental question and one that needs to be answered in order to proceed with the investigations, because it has consequences for those investigations. The term *sociotechnical system* is rather general. It is thus necessary to understand the sociotechnical system in the context of the case and the objective. The treatment of this research question also contributes to defining the empirical field, which is an essential delimitation for the writer and a useful introduction for the reader.

The further exploration of the main objective leads us to studying work processes 'in the wild', i.e. the informal and contingent structures of work that do not necessarily reflect formal work procedures which are established in procedures or governing documents. This theme also emphasises the orientation of the study towards the organisation not as a representation, but as real, empirical activities. Whereas safety of sociotechnical work is often sought to be consolidated through formal mechanisms, the contingent nature of sociotechnical work will challenge the relevance of the formal representations. The second aspect of the main objective thus addresses the ongoing, unplanned work within a sociotechnical system. It explores the mechanisms of such work that may contribute to operational safety, by asking the following question: *What characterises the informal coordination of*

---

to *what we know about this world and how the knowledge is produced.* In a constructivist perspective, however, these two aspects are two sides of the same coin. The use of these terms in this book will therefore be based on other and more subtle conditions and connotations than what is traditionally the case. Latour (1987, 1993, 2005) is a good source for those who want to read more about this constructivist approach.

4   Importantly, that is not to say that the first three research questions only enter into the discussion as intermediaries on the way to answering the fourth research question. Each of them offers important, independent insights.

work in the occurrence of unexpected events, and how may the significance of such work for operational safety be affected by the introduction of new technologies?

Collaboration between people with different professional backgrounds and operational goals is a central part of the work processes in drilling operations. The third aspect of the study's main objective concerns the role of shared understanding of ongoing operations across the disciplines for the safe accomplishment of the operations. In the research literature, the significance of shared understanding is repeatedly underscored, whereas one seldom finds the ontological conditions and consequences of such commonality being subject to critical examination. This third aspect resembles the first aspect in that it addresses the basic nature of a phenomenon that is often taken for granted and treated in a general and non-falsifiable manner. The third question that is asked is: *What is the role of shared understanding in the interpretation of data in multidisciplinary teams?*

Having shed light on three distinct, specific aspects of the main objective, the study is rounded off by addressing the main objective at a more generic level. Whereas the first three research questions are elaborated on by referring to cases, the fourth elevates the discussion by introducing a broader spectrum of actors into the discussion and by having them explicitly reflect upon the study's main objective. In the elaboration of this research question we can hear the echoes from the first three. Having elaborated on the nature of sociotechnical systems and the interpretation and coordination of work among the systems' heterogeneous actors, an understanding of the outcome of sociotechnical work is established as not having resulted from the properties of predetermined actors, nor from any generic model of safety or collaboration, but from the diverse actors' situated and non-standardised solutions to the continuous challenges of sociotechnical systems. The fact that challenges are unpredictable and solutions are situated and non-standardised does not mean that sociotechnical systems are impossible to control. It indicates, however, that our representations should reflect the dynamics and uncertainties of the sociotechnical system rather than generic models that are fit all purposes. Thus, the fourth and last research question is: *How will Integrated Operations influence the conditions for safety of drilling operations?* This question is asked in order to identify central challenges of drilling operations, on which new technologies may have an impact.

The subject of this book is loaded with controversy. The book does not, however, reflect any viewpoint on these controversies. That is not to say that the study is apolitical. On the contrary, it is fundamentally political – as opposed to ideological – in the sense that it represents an attempt to account for the matters of concern without assuming a priori the actors, processes and powers at play.[5] Thus, at the starting point of the inquiry the actors who will populate the descriptions are not yet identified, and a social theory that explains their behaviour and the power relations between them is not adopted. Through observation, interviews

---

5 See Latour's (2004b) elaboration of the concept due process for a source of inspiration on this perspective.

and case studies the actors' own work and their own articulation of it are explored. This methodology constitutes the main research strategy of the study and arguing strategy of this book.

It would not have been possible to write this book without having personal experience from the offshore domain. Three years' experience as a professional mud-logging geologist is reflected in the way the book's empirical material is chosen and elaborated. A thread through the book is the presentation of work as it is actually done, in contrast to how it is prescribed by flowcharts and governing documentation. Some of the empirical material thus represents collaborative work that an outsider would have a hard time getting access to, both physically and intellectually. The physical aspect involves both the offshore and onshore locations. While it may be very difficult for a researcher to get permission to study the work on the drill floor, in the driller's cabin and in the office of the data engineer and mud-logging geologist on an offshore rig in the first place, for an outsider to understand the technical aspects of the work from the perspective of those who perform it is also challenging. While getting access to the onshore locations is definitely easier than getting access to the offshore locations, it might be even more difficult to understand the higher-level work performed there, not to mention grasping the details of the offshore work from that onshore position.

These aspects of research work may appear as merely practical issues – and they are indeed highly practical – but they are more than that; they also touch upon central methodological aspects of research activities: from which position is the researcher observing the field, and in which position are they to produce descriptions that are not only meaningful to the research community, but also to the field itself? I find this important not only as a generic question. It has also been so influential to the work this book is based upon that it is granted further elaboration in Sections 1.3 and 1.4.

1.2 A 'Positive' Safety Approach

While safety may be thought of as *successful work* on one hand and as *absence of failures* on the other, the two perspectives do not belong to different realms – their natures are not even qualitatively different: 'Failure is the flip side of success, and therefore a normal phenomenon' according to Hollnagel et al. (2006: 14), with strong connotations to Perrow's (1984) notion of *Normal Accidents*. Both Resilience Engineering (Hollnagel et al., 2008, 2006; Nemeth et al., 2009) and the High reliability Organisations (HRO) approach (La Porte and Consolini, 1991; Weick and Roberts, 1993) provide theoretical frameworks oriented at accounting for normal successful work[6] and at describing the characteristics of the organisations that undertake it. According to these traditions, successful

---

6   Note that resilience engineering also offers frameworks that may be used to account for failures, such as the Functional Resonance Accident Model (FRAM).

outcomes do not presuppose nor reflect absence of risks. Rather, successful outcomes reflect how risky and intractable sociotechnical systems are managed by an organisation that is able to 'respond to events, to monitor on-going developments, to anticipate future threats and opportunities, and to learn from past failures and successes' (Hollnagel et al., 2011). Another way of characterising organisations that successfully manage high-risk activities is to emphasise a set of generic properties, i.e. such organisations' preoccupation with failure, reluctance to simplify, sensitivity to operations, commitment to resilience, and deference to expertise (Weick et al., 1999). Yet another set of features is the organisational redundancy of such organisations, combined with their ability to reconfigure spontaneously from a centralised to a decentralised structure in order to mobilise the relevant expertise during demanding operations or in crisis situations (La Porte and Consolini, 1991).

In the traditions of Resilience Engineering and HRO, safety is seen as resulting from continuous adjustments to cope with varying conditions. It is usual in both approaches to refer to safety as a *dynamic non-event*. This term was originally developed by Weick (1987) to describe *reliability*, but has since been widely used also in the context of safety (e.g. by Hollnagel et al., 2006; Reason, 1998). Safety is described as *dynamic* in the sense that it is 'preserved by timely human adjustments' (Reason, 2000: 770), and it is described as a *non-event* 'because successful outcomes rarely call attention to themselves' (Reason, 2000: 770).

That safety is a dynamic event is not difficult to agree on. Indeed, most researchers exploring work practices in safety-critical industries report that plans are 'resources for action' (Suchman, 1988: 313) rather than determinants of action, implying that the characteristics of safe outcomes are work practices of articulation, improvisation and adaptation rather than compliance. The dynamics of safety are reflected in Suchman's statement that 'every course of action depends in essential ways upon its material and social circumstances' (Suchman, 2007: 70).

Safety is thus definitely dynamic.[7] But is it a *non-event*, in the sense that 'successful outcomes rarely call attention to themselves' (Reason, 2000: 770) or that 'nothing is happening' (Weick, 1987: 118)? It may be, if we choose to displace safety from where we located it a moment ago, in the ongoing work processes, to the final result, located somewhere along an axis of time or work processes. Or it may be, if we choose to abandon our description of the dynamic adjustments involved in the operators' work, the 'human variability in the shape of compensations and adaptations to changing events represents one of the system's most important safeguards' (Reason, 2000: 770). But why such a displacement? Does it not make more sense to maintain the view on safety as an *integrated part* of the work processes as such, not as a post hoc *result* of them? And if so, does it not seem strange to emphasise that safety – understood as a fundamental quality

---

7 Although in extremely reliable and more or less closed systems where the main work of the operators is to monitor the processes, not to intervene, one may talk of safety as a static, non-event.

of work – is so invisible[8] that it may be called a non-event, instead of looking for methods for making this work visible *for others than those who perform it*? There is no doubt that when an operator is controlling a process dynamically by timely, minute, human adjustments, those adjustments are responses to the feedback from the process. And that feedback, that outcome, is far from nothing to the operator – it is his whole world, which he acts upon!

In line with that, in this book safety will be treated as a *dynamic event*, that is, a phenomenon that is characterised by continuous actions, adjustments and adaptations that may seem indistinguishable from the primary tasks of the drilling operations. This book is not about the meta-work of safety; traditional HSE work such as near-miss registrations, safety culture campaigns, HSE performance indicators and other activities whose monitoring primarily is the responsibility of the HSE engineers do thus *not* fall within the scope of the book. What *do* fall within the scope are all those activities that are performed in order to construct the offshore wells, indeed under risky conditions, but still activities that more often are thought of as belonging to the realm of production and efficiency than to the realm of safety. That is not to say that safety and efficiency is the *same* phenomenon, but that they are different aspects of the same phenomena.

In other words – and rather than hair-splitting this is one of the main messages of this book – viewing safety as a dynamic non-event may be misleading and may obscure the activities that actually produce the safety, thus inhibiting the possibility of strengthening and rendering these activities more resilient. On the contrary, it is the conception and understanding of safety as *dynamic events* that have been the point of departure and the source of inspiration for the work that is presented in this book, and this has been an indispensable prerequisite for asking the questions that are asked, and for pursuing their answers.

1.3 Work and Fieldwork – Getting Access to the Field

The main locus for the fieldwork was an onshore operation centre in an international oil and gas company. Much of the observation studies, interviews and document studies were undertaken here. The empirical field also included support centres, service companies and research and development institutions.

One of the more challenging parts of fieldwork in a commercial company will often be gaining access to the field in the first place. The company will have to consider several issues before letting the researcher inside: Will he be in the way? What access to sensitive information should he be granted? How much time and resources can the company afford to allocate to him in order to provide him with what he wants? What is in it for the company? Well aware of these issues, the researcher must also consider his strategy to satisfy the basic conditions of the

---

8   Invisible work is definitely an adequate description, but only in the sense *invisible to others* (see Suchman, 1995).

company and to make it tempting for them to let him in, without giving up the central goals of the research and without compromising its independence.

Having been provided with a key card and office accommodation at the company's research centre,[9] my professional background from the offshore industry was of great importance for establishing the necessary relations and agreements with specific persons and groups to interview and observe. Three key words sums up the most decisive conditions in connection with this: language, need for follow-up and trust.

Having the professional experience and thus speaking the tribal language of the potential informants made it easier to explain my intentions in a way that was understandable and relevant to them. This also made it possible to appear as an insider who would be able to perform my tasks without depending on the company taking any initiatives towards me and allocating resources specifically for following me up. The familiarity with the industry from the inside also helped in building trust between me and the potential informants I visited. This is not to say that they were confident that I would not reveal conditions that were unfavourable for them and that I was therefore considered harmless, but rather that they trusted that I was competent to reveal issues with relevance for their work.

1.4 Work and Fieldwork – Insider and Outsider Perspectives

In retrospect, one could say that the fieldwork for this study started back in 1997, when I was employed as an offshore mud-logging geologist[10] for an international service company. At that time I had no idea that the three years I spent in the company would constitute an important contribution to a study that was to start more than ten years later. At that time, I simply did the job I was expected to do, reflecting on safety and performance from the professional view of a mud-logging geologist, not from an academic researcher's perspective.

During the following three years on the Norwegian continental shelf, I worked under 4 different oil and gas companies, and at 12 different drilling rigs. I attended courses in mud-logging, formation evaluation, primary well control, overpressure, d'exponent[11] and general drilling safety, in addition to general survival training. Considered as fieldwork, it was an invaluable experience. What I learned about drilling operations during those years I could not have dreamed of learning from

---

9   This access was provided by an influential employee in the main company of the study. This person is well respected both by the industry and in academic circles, and the goodwill of this gatekeeper was invaluable for the project.

10   The job of a mud-logging geologist is to examine and analyse the bits of rock (cuttings) that are brought to the surface during drilling, and to create detailed lithology plots of the borehole.

11   The d'exponent is a parameter that is used in formation pore pressure evaluation while drilling.

conventional fieldwork alone. Not only was I able to develop a professional vision (Goodwin, 1994) in the field, I was also given insight into work processes and situations that would be difficult to explore as an outsider with access merely to the tools of observation, interviews and literature review.[12] As an insider I also had access to the discourses that informants are perhaps not eager to discuss with researchers; such discourses could for example include aspects of work that are inconsistent with rules or normal procedures, but still are aspects of the efficiency–thoroughness trade-off (Hollnagel, 2009) that characterise much work.[13]

This insider position was re-established when I again entered the field in 2008. The difference now was that I was able to see the field from another perspective, and hence could profit from a double description, a concept that metaphorically compares with the depth vision made possible by binocular vision (see Bateson, 1982). This has been an important premise for the study. Depending on the perspective, there are numerous possible descriptions and elaborations of the same setting and situation, and the goal of this study was to explain the work of drilling operations in a way that seemed relevant to drilling professionals and researchers of work and safety at the same time, and *in the same way*. Although the advantages and disadvantages of being an insider or an outsider – a *stranger* – are difficult to sort out conclusively (for a more elaborate discussion of this, see Latour and Woolgar, 1986; Lynch, 1982), the 'double professional vision' was nevertheless essential in this specific study.

The empirical focus of science and technology studies that was adopted in the study requires the researcher to understand the work of the informants on their own premises to avoid ascribing to their work motives and powers that they do not recognise or acknowledge themselves. Nevertheless, to avoid producing explanations which are nothing more than simple reproductions of the informants' own accounts, it is necessary to be able to look up and see the work in a wider context than that which each of the actors may easily perceive themselves – to follow the traces of the networks outside the radius of action of the local actors. This is where the professional approach of the social scientist is vital. In summary, these aspects of the study contributed to narrowing the gap between

---

12   This is not to say that a comprehensive ethnographic study performed by researchers without the field profession cannot capture important features of professional work; many such studies have been undertaken with great success (see e.g. Hutchins, 1995a; Latour and Woolgar, 1986; Orr, 1996; Traweek, 1992). To obtain a professional vision, however, requires more time and resources than I was granted for this study.

13   One example of this could be the extension of the 12¼" section into the reservoir (see Figure 2.2). This is a practice that the petroleum safety authorities do not recommend due to the kick risk. Another example is the main issue in the same case – the depths measurement discrepancies. This was not mentioned at all in the final well report. The significance of these issues would be very difficult to evaluate – if the issues would be noticed at all – for an observer without in-depth knowledge of the field.

work and fieldwork, hence also hopefully the gap between the informants' and the researcher's comprehension, as it appeared towards the end of the study.

Exploring this gap actually turned out to be an important part of the fieldwork and analysis work as such. By closing the loop through bringing empirical data and preliminary analyses back to the informants and other professionals from the industry for secondary discussions, not only the researcher but also potentially the informants could develop their understanding. This double hermeneutics (Giddens, 1987) also functioned as a motivating process and a corrective in times when both I and the empirical data were 'carried away' by theories whose connection to real life was not so obvious.

Chapter 2
Drilling in Action: Two Short Stories

In this chapter the reader is invited to be my co-observer of the work of the onshore personnel supporting the offshore operations. Through two cases we will get a unique insight into the nitty-gritty details of normal operations, and we will learn how contingencies are a part of normal work and are managed with great pragmatism. The cases are presented in brief here so that the reader can get a feeling of where we are heading and to adjust the modus of thought as early as possible. In Chapters 8 and 9 the cases will be subject to deeper analysis.

2.1 Collaborative Work in a Mud-loss Situation

In the morning, the expert centre receives a phone call from an onshore rig team experiencing loss of drilling fluid. The rig crew does not know where the losses occur; whether it is to a specific geological formation in the well, through a crack in the well or even topside. The main challenge is to determine the location and cause of the losses observed. The role of the mud[1] in maintaining well integrity implies that lost control over the mud circulation might have very severe consequences. However, because the loss rate is low and tolerable, the situation is by no means considered critical at the time; consequently, the rig crew continues to drill while searching for the explanation. Figure 2.1 depicts the flow system of the drilling fluid, and the arrows indicate the flow path.

The rest of the members of the expert centre are immediately mustered, a video conference is established with the rig crew and real-time data are displayed on a large screen, which is also visible to the onshore rig team. The situation is discussed without reaching a conclusion before adjourning the meeting. This allows the expert centre to further evaluate the situation with the information available before returning a recommendation to the rig team by the agreed time. What happens from this point onwards is that different interpretations are elaborated and explored. The interpretations do not arise from a standardised checklist. Rather, they are formulated by different actors situated in different places in the drilling organisation, by means of the available information, knowledge, experience, discussions and different tools for presenting and mediating the information.

A highly detailed account of all the different interpretations and their technical origins is beyond the scope of this chapter. For the purposes of this case study, four identified interpretations, or conjectures, are identified and discussed:

---

1 Mud is sometimes referred to as drilling fluid.

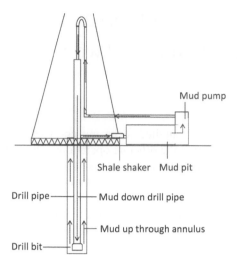

Figure 2.1   The circulation of drilling fluid[2]

First Conjecture

One of the members in the expert centre observes that a very similar well was drilled some weeks before, in close proximity to the well under consideration. That work was performed by the same drilling rig, and they had suffered mud losses similar to those now under investigation. Another member of the team, responsible for facilitating access to real-time and historical data to the rest of the team, retrieves the relevant data for the event from a real-time data web portal. A third member of the team has previously developed a means of presenting this information in a manner specific to investigating mud consumption and loss during drilling; the template was prepared using Excel, which is a tool common to all actors involved in this case, and was designed specifically to be intuitive in use by any member of the team. The objective of this idiosyncratic presentation was to highlight potentially obscure data related to mud characteristics and volume trends by means of a standard format in order to expedite comparison between different cases, thus allowing common causalities to be identified across multiple cases and drilling locations and by various actors.

---

2   The mud serves several different functions in a drilling operation. First, it transports drilled cuttings from the bottom of the well to the surface. If this transport fails, the drill string might get stuck because of accumulation of cuttings in the well. At the surface, the cuttings are separated from the mud at vibrating sieves, called shale shakers. Second, the mud functions as a barrier against gases and fluids in the formations. Other functions involve using mud pulses to transport signals between technical equipment down-hole and control and interpretation equipment topside.

Critical assessment of mud-consumption plots from both wells is thus undertaken. Although similar trends that support the notion of a common mechanism are observed, in this case a seepage loss to a porous or fractured geological formation, the mud volume trends depart from the anticipated volume trends at different depths and in neighbouring geological formations, undermining the case for a common cause.

Second Conjecture

Judging by the relevant reports, the mud system in use on the current operation is, qualitatively, very different from the mud system used in the previous well. This makes direct comparison of operating parameters and the down-hole environment less meaningful. The members of the expert centre perform calculations to determine whether the new mud is, in fact, too dense, thus creating a pressure in the hydrostatic column that exceeds the maximum pressure that the rock strength of the geological formation theoretically allows.

This investigation concludes that the mud density is comfortably within the allowable limits and there is thus no clear indication that the formation has been fractured, allowing mud to flow from the well into the surrounding rock.

Third Conjecture

The expert centre now opens a video conference with the onshore rig team's geologists, who are located in another city, in order to discuss the geological properties of the formation and what interactions might be possible between the formation and the mud. The rationale is to determine whether any undesirable chemical reactions between formation and mud is plausible and, if so, what influence these might have upon the calculations of hydrostatic pressure referred to previously. If they can identify any plausible chemical reactions that may have reduced local formation strength, the preconditions upon which allowable mud density had been determined may have been rendered invalid. In that case, the road map for the drilling operations, or alternatively the properties of the mud system, will then have to be adjusted accordingly.

Given that the mud system employed is quite new, little operational experience with it exists and the discussion is unable to substantiate the possibility of any chemical reaction that could lead to instability.

Conclusive Interpretation

Later, at the time agreed during the morning video conference, two members of the expert centre, who happen to be based in a building nearby, visit the rig team in order to present the likely cause of the mud losses and the recommended remedial action to take. The result of the discussions is explained to me the following day. While investigations were underway, the mud logger had inspected the shakers, the function of which is to separate rock cuttings from the drilling fluid, to dispose of the cuttings and recycle the mud. It transpired that the mud loss was due to mud passing over the shakers and not to the formation as had been envisaged. The losses were found by the rig crew themselves to be an unfortunate and unforeseen combination of properties of the mud and the cuttings. The shakers were not separating all of the mud from the rock cuttings effectively, owing to the mud's propensity to adhere to the surface of the rock. Having identified the nature of the mud loss, the organisation was in a position to determine and affect a solution.

2.2 Divergent Depth Measurements and the Role of Shared Understanding

Contingencies appear frequently in drilling operations. One challenge faced by the involved actors is thus to relate to new information that requires interpretation and negotiation and to revise plans and actions accordingly. The complexity of this case makes a brief introduction necessary. Before the case is described, let us examine some characteristics of drilling operations that are made current in the case.

Cooperation between the different disciplines involved in the drilling operations involves extensive articulation work.[3] The division of labour also implies a division of perspectives, goals and performance measures. Different disciplines will have different goals and consequently different performance evaluation criteria. The collective and complex task of drilling and production is characterised by one informant by the 'inherent dilemmas of the petroleum industry'. The decision-making involves trade-offs between efficiency and thoroughness (Hollnagel, 2009), between conflicting goals among the disciplines and between short- and long-term perspectives.[4] A common view is that drilling engineers have a short time horizon; their job is finished when the well is drilled. The success criteria are that they manage to drill the well so that it is technically optimal, with as few and gentle curves as possible, at a safe distance from other wells in the area, with little down-time and no accidents. Reservoir engineers, on the other hand, have a longer time horizon for their work. The positioning of the well in the reservoir has a significant effect on the long-term drainage of the reservoir. Hence, whereas a decision regarding the well path in the reservoir is a question of technical possibilities for the drilling engineers, it is a question of long-term hydrocarbon

---

3   As we will see in Chapter 8, articulation work is a way of conceptualising the work described in the previous case.

4   Although other actors might be involved in issues like the one discussed in this chapter, only two groups of actors will be considered in this discussion to make the argumentation clear and conceptual. The actors belong to the drilling and well department and the petroleum technology department and will in the following be labelled drilling engineers and reservoir engineers. The onshore rig team, which was the main locus of the fieldwork, consists mainly of drilling engineers.

flow, production and expected profit for the reservoir engineers. Consider the following statement from one informant:

> To the drilling engineers, a project may be successful if the well is drilled without collisions with other wells in the area and without experiencing any serious well-control issues. To the reservoir engineers, on the other hand, a successful well is one that is located perfectly in the reservoir, as one well out of many, so that the total, long-term production from the reservoir will be as high as possible. (Reservoir Engineer)

Another informant addresses the issue by referring to specific, strategic trade-offs between drilling efficiency and field knowledge development: a geologist might advise reducing the rate of penetration through a specific formation to sample the formation thoroughly. This could be a strategy to understand and improve the models for the larger field, and might not necessarily be of any value to the ongoing well project. In such a situation, the extra drilling time will be accounted for in the drilling budget, and the drilling engineers might therefore be reluctant to choose such a strategy because it represents an expense to their budget without any prospects for profits in the same budget.

Despite these conflicting interests, there is no way any of the disciplines can obtain their goals without collaborating closely with each other. The disciplines collaborate in every phase of a project, from planning to completion. They are also aware that one discipline's goal achievement is worthless if it means that the other discipline does not reach *its* goal. The different goals and perspectives are typically made current when the agreed drilling programme is challenged by contingencies. The case description below is a case of such contingencies, and it illustrates a central point: the different ways of relating to and accounting for the discrepancies reflect more than simply different goals; they reflect different epistemological approaches that do not easily integrate to support the different aspects of shared understanding. Be aware that the case is not an extraordinary event for the involved actors, and the handling of the situation is described as ordinary work.[5]

Case Description: Depth Measurement Discrepancies

The rig team that was studied is located in one of the operating company's onshore operations centres. The centre operates a number of offshore fields, and each field

---

5 To the extent that any work in drilling operations can be described as ordinary; the contingent, non-standardised nature of drilling operations, where the underground formations are never identical in two operations, and are never fully known, implies that every ordinary work process still is unique. That this case is not a special incident without generic implications is illustrated by the fact that it is not mentioned in the final well report, which is a summary and a learning document from the drilling operation.

is represented by one rig team. The responsibilities of the teams are to produce drilling programmes and to follow up their execution.

A well is drilled in several sections. Each section has a different diameter: largest for the upper section and smallest in the last, deepest section, which penetrates the reservoir. A typical sequence of sections' diameters is 36" (inches), 26", 17½", 12¼" and 8½". The transition between sections is often determined by the boundary between two geological layers. Such boundaries therefore play a role both in the drilling process and in the later production stage, because the boundaries define areas from which oil can be produced. Other types of boundaries that are used for navigation are the interfaces between fluid and gas phases such as the gas/oil contact and the oil/water contact in the reservoir. It was uncertainty in connection with the depth measurement of such boundaries during a drilling operation that formed the point of departure for the case.[6] The schematic presentation of the case offered in Figure 2.2 is a useful reference for the following case description.

The advice in the drilling programme was to place the 8½" well section horizontally in the reservoir in a position relative to the gas/oil contact and the oil/water contact that would ensure production of oil without influx of gas or water.[7] To do that, it was crucial to determine the precise depth of either the gas/oil contact or the oil/water contact. These depths were found by logging the pressure gradient and the resistivity of the fluid along the well. A break in the pressure gradient will indicate the contact zone between different fluids. An accurate determination of the depth of top Garn formation (see Figure 2.2), was also important because it defined the top of the reservoir.

In connection with depth measurements, it should be noted that there are two different types of depth values; measured depth and vertical depth. Measured depths are calculated as the sum of every joint making up the drill string from surface to the drill bit. The composition of the drill string is listed in a manually produced paper or computer file, the tally. Vertical depths are then calculated on the basis of the measured depth and the curvature of the well path.

In the case, the reservoir was first penetrated with a 12¼" drill bit. The Garn formation was identified and the vertical depth of the formation top was determined through a combination of measurement and calculation. Top Garn was identified 14½ metres deeper than forecast. The gas/oil contact was not found.

---

6   Several situations that appeared during the field study could have served to illustrate the point of this case. The actual case was chosen partly because it served well to illustrate the different aspects of shared understanding, partly for practical reasons; it elapsed during a limited period of time and involved actors whose contributions were possible to get an overview of during the observation and the following interviews.

7   We now need to be a bit technical, and this may be challenging for the layman. There is no way around this, unfortunately, but the dictionary at http://www.glossary.oilfield.slb.com may offer good help. This case and the later discussion of it will be far more interesting if the terminology is well understood.

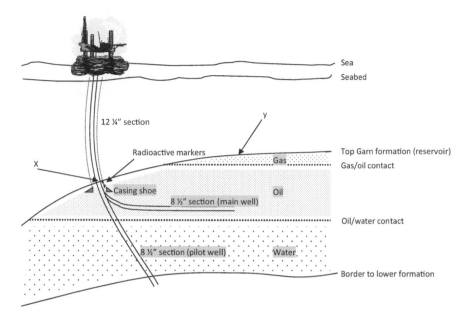

**Figure 2.2   Placing the well in the reservoir (the figure is conceptual and
not correct in scale or details)**

According to the informants, this could be the case because the boundary between
top Garn and the overburden formation was sloping, and that the penetration of
Garn was at location x and not at, for example, location y, as shown in Figure 2.2.[8]
It was therefore decided to continue drilling a *pilot well* (see Figure 2.2) through
the oil/water contact with an 8½" drill bit, and then use the oil/water contact and
not the gas/oil contact as a depth reference point. Before this could be done, a
casing needed to be run to secure the 12¼" hole. At known positions inside this
casing, two radioactive markers were placed to function as easily detectable depth
reference points for the next drilling section.

When the drill string subsequently was run into the hole and through the casing
with an 8½" drill bit, some unexpected depth discrepancies were observed: top
Garn was identified 4.5 metres deeper in the 8½ section than in the 12¼ section.
In addition, when the depth of the radioactive markers was logged, there was a
discrepancy of 1.9 and 0.5 metres respectively between these measurements and
those performed in the 12¼ section. Although it was the oil/water contact and
not the top Garn or the radioactive markers that was the reference point to be

---

8   There was no gas/oil contact present at that location. Due to different specific
gravities, the fluids in the reservoir will migrate so that gas gets on top, oil in the middle and
water on bottom. The border between Garn and the overburden formations is impermeable.
Figure 2.2 shows the position of the well and of the gas/oil contact.

used when later kicking off laterally for the main well (again, see Figure 2.2), this discrepancy raised a dilemma: which measurements were correct – those measured in the 12¼" section or those of the 8½"? And if the 8½" pilot well measurements were unreliable, how could they serve as a reference for the subsequent main bore? As we shall see, this last question was related to differently by the different disciplines in the organisation.

Apart from this main depth issue, there was another topic that indirectly related to the main issue and that had some of the same characteristics with respect to how it was related to by the different disciplines. In a video conference between the onshore rig team and the offshore rig crew, the offshore operational geologist argued that they should take more pressure measurements in the reservoir than initially planned for. The rationale for this was to get a better understanding of the structure and the geology of the reservoir and to be able to make more accurate model descriptions of the boundaries between the gas, oil and water for the larger field. The drilling engineers would rather avoid this, because such operations are time-consuming and increase the risk of getting stuck due to the required standstill of the pipe during such measurements.

There were thus two issues that the drilling organisation had to deal with. First, there was the question of the different depth measurements and the uncertainty of their reliability. This will be called issue A. The second issue concerned the number of pressure measurements to be taken, and this will subsequently be called issue B.

Possible Explanations Elaborated by the Participants

Issue A was the most troublesome, because there were so many uncertainties attached to it. There were many possible sources of error that challenged the reliability of any interpretation of the state of affairs. The issue was first considered in the morning meeting, which was conducted as a video conference between the rig and the onshore team. After the meeting, the reservoir engineers and the drilling engineers discussed the issue separately. The issue was again brought up at the morning meeting the following day. The alternative explanations that were discussed in this meeting will be briefly reviewed below.

During the meeting, six main potential sources of error causing the measurement discrepancy were identified. The first potential source was an error in the tally, where the lengths of all components of the drill string are listed. If for some reason the tally did not correspond to the actual make-up of the drill string, the measured depth would be erroneous. The second potential source was the radioactive markers which might have loosened and been pushed upwards when the casing was run downwards. The third potential source was the tide which, according to the informants, was not usually taken into account during operations of this kind.[9]

---

9   The installation was a floating rig, hence the level of the tide has significance for the depth measurements.

If the depth measurements were not correlated to the corresponding level of the tide for the two measurements, the variance could amount to 1½ to 2 metres. The fourth potential source was a possible difference in the stretching of the drill string due to the string's different weight in the respective sections. In the 12¼" section, during the casing running, the stretching of the drill string would be negligible. In the 8½" section, on the other hand, the drill string was run into an open hole and would be much more stretched. This would theoretically lead to a too small depth measurement. One of the informants made it clear after the meeting that they usually not did account for such variations. The fifth potential source was the identification of the top Garn formation in the two runs, which was based on resistivity measurements in the formation and was subject to human interpretation of resistivity logs. The sixth potential source was the amount and the effect of ballast on the rig during the two runs. It was concluded that ballast could lower the rig by as much as 2–3 metres in the sea. The participants were not sure whether any differences in ballast had been accounted for in the two measurements.

Issue B was not subject to as much discussion as issue A. In the second morning meeting, the offshore geologist requested more pressure points to be taken than they initially had agreed upon in the reservoir section. No decision was made at the time, but the geologist was asked to reconsider whether this was really necessary. The issue was elaborated on by some of the rig team members during lunch the same day. One drilling engineer commented that every extra pressure point would take half an hour and cost approximately US$8,000. Another engineer noted that the price of such measurements might not be the most important counter-argument. The increased chance of getting stuck due to the standstill of the pipe when the pressure points were being taken would be more serious. He concluded that 'we don't want to spend time on things we don't need'. A third engineer added, humorously: 'Things *we* don't need, no', addressing with obvious self-irony the different requirements of the different disciplines and the difficulties of fulfilling them all at the same time.

The depth discrepancy issue was never fully solved. The possible explanations were discussed and attempts were made to combine them to see if they could sum up to the total difference between the measurements. However, when the drilling operation continued into the 8½" section, there existed only different theories and no unambiguous solution to the case. It was agreed to proceed without any absolute references of depth, to place the well relative to the oil/water contact and to be attentive to any clues that might appear on the way.

2.3 Methodological Aspects

One striking feature of these two cases is the complicated[10] organisational and operational arrangements that characterise the domain of drilling and well. Later, we shall learn from these detailed accounts that some of the success factors of Integrated Operations that are often taken for granted may be problematic and even counterproductive if the goal is increased safety. But these conclusions are not obvious, because they rest upon some epistemological conditions that are parts of the methodological framework adopted by the researchers. The methodological framework guiding the research that is presented in this book must therefore be accounted for before the analyses of the cases are carried forward. In Chapters 7–10 we shall see how this framework and this epistemology both constitute constraints and offer opportunities for the collection and analysis of empirical material. But before that it may be useful to elaborate on the concept of Integrated Operations and, in a very simplified manner, introduce some core concepts of drilling operations. This may help readers who have no training or experience with the drilling domain by providing a context for the subsequent reading of the book, especially the more empirical parts.

---

10   I intentionally use the word *complicated* here, instead of *complex*, which is often used in similar contexts. Complexity is a notation with ambiguous connotations, and I will illustrate the importance of precision in the use of this word in my later elaboration of sociotechnical systems in Chapter 7.

Chapter 3
A Brief Introduction to Integrated Operations

If we now let the two cases introduced in Chapter 2 rest for a while and spend some time on understanding what exactly is meant by Integrated Operations (IO), we will be better equipped to understand how IO fits into and changes a drilling organisation and its operations. Later we shall use this knowledge to work out how the monitoring, diagnosing and automating tools may influence operations in general and the cases specifically.

There is no single definition that captures all the aspects and nuances ascribed to IO by the different stakeholders in the industry and the research communities. One generic working definition that most stakeholders may agree on is that IO denotes *the integration of people, work processes and technology to increase the quality and speed of decisions and execution*, and that this is enabled by the *use of ubiquitous real-time data, collaborative techniques and multiple expertise across disciplines, organizations and geographical locations.*[1] The problem with such a definition, however, is that it is so general that it doesn't draw any boundary around the concept to decide what *are* Integrated Operations and what are *not* Integrated Operations; it doesn't exclude anything from being labelled Integrated Operations. And couldn't this definition simply refer to change processes that a range of other industries have already undergone, thus turning IO into uninteresting, yesterday's news? And truly, in much IO and safety research parallels are frequently drawn to other domains that experience similar challenges of complexity, coordination, cooperation and communication such as air traffic management, nuclear and military industries.

The advantage of such a generic definition, on the other hand, is that it motivates a variety of research and research perspectives on the petroleum industry, and hence invites the research community to operationalise and shape Integrated Operations – actually developing it – in accordance with scientific perspectives on, for example, safety. Much IO research would perhaps not have taken place with a narrower definition, and this in turn would surely deprive the petroleum industry and other comparable industries for many initiatives with great value potential in terms of safety, efficiency and sustainability. The variety of IO research, here represented by only a small selection of titles all of which fit into the IO definition, illustrates this: *Automatic real-time drilling supervision, simulation, 3D visualization, and diagnosis on Ekofisk* (Rommetveit et al., 2008a); *Investigation of feasibility and potential for sub-surface imaging using wired drillpipe in connection with seismic-while-drilling* (Helset et al., 2011);

---

1   Based on the definition from http://www.iocenter.no/doku.php.

Knowledge markets and collective learning: designing hybrid arenas for learning oriented collaboration (Bremdal and Korsvold, 2012); *Between and beyond data: how analogue field experience informs the interpretation of remote data sources in petroleum reservoir geology* (Almklov and Hepsø, 2011); *A Lagrangian-barrier function for adjoint state constraints optimization of oil reservoirs water flooding* (Suwartadi et al., 2010); *Technical health of a system – in the context of condition based maintenance* (Vaidya and Rausand, 2010); *Shared collaboration surfaces to support adequate team decision processes in an Integrated Operations setting* (Kaarstad and Rindahl, 2011); *Demonstration of a research prototype of a collaborative planning tool for use in offshore petroleum operations* (Veland and Andresen, 2011); *The influence on organizational accident risk by Integrated Operations in the petroleum industry* (Grøtan et al., 2010); *An investigation of resilience in complex socio-technical systems to improve safety and continuity in Integrated Operations* (Johnsen, 2012).

Rosendahl and Hepsø (2013) and Albrechtsen and Besnard (2013) offer excellent overviews over the current research status in the IO field.

3.1 Terminology

The use of the term Integrated Operations throughout this book does not reflect a view of this concept as superior or preferable to the many other concepts that exist internationally in the oil and gas industry. Whereas IO is an agreed term in the Norwegian part of the industry, other companies and other sectors have adopted different terms with slightly different connotations. Hence field of the future, smart fields and iField are comparable terms referring to comparable goals and comparable means, as are digital oil fields, e-field, intelligent energy and digital energy. It is important to remember, however, that the relatively brief formulations of these initiatives are far from capable of synthesising the substance of what they denote. What this book addresses is not the uniqueness of Integrated Operations as such, as compared to other concepts, but rather how the impact of new tools and work processes related to information and communication technologies may be studied and understood, whether these are defined as Integrated Operations, field of the future, smart fields or *any other related concepts* for improving safety and efficiency in the oil and gas industry. To fulfil such an ambition it is necessary to get under the surface of the theoretical and visionary definitions and study what is actually going on the field and how the changes do and will interfere with this work in practice. And this practical world can be quite ruthless to theories and visions …

3.2 Historicity

From an original denotation for processes and tools for effective real-time utilisation of increased amounts of data, IO has gradually developed into a rather generic term covering most aspects of the oil and gas industry that involves information, communication and cooperation. Integrated Operations represents an operating mode that most oil companies have decided to adopt and develop in all parts of the industry; drilling and well construction, reservoir management and production optimisation, operation and maintenance, logistics and licence administration.

Integrated Operations can be characterised both as a conscious *decision of implementation* and as a gradually *emerging development* of the operations. The understanding of IO as a *decision* can be traced back to the early writings on the topic, where the recommendations are to *implement* IO in order to increase the efficiency and safety of the industry (OLF, 2003, 2005). The other understanding of Integrated Operations is that it is a part of a continuous industrial evolution that has eventually been given a name. Such a framing of the change process makes it fit well into the pattern of 'efficiency leaps' that has characterised the development of the industry during the last few decades. In the mid/late 1980s, a range of new technologies were successfully introduced; improved 2D seismic, the top drive,[2] new logging while drilling (LWD) techniques and alternating water and gas injection WAG, among others. Then, in the mid/late 1990s, a second leap introduced 3D seismic, horizontal drilling and geosteering.[3] Both of these leaps brought about reduced costs of development and operation, improved oil recovery, reduced oil spill and improved safety. Integrated Operations is expected to represent a third leap, and the effects of the first and second leaps are thought to be further strengthened (OLF, 2003).

Although early descriptions of IO characterised it as an ICT development, it was also ascribed a potential to render possible a whole new range of technologies and work processes if it was combined with new oil and gas technologies (OLF, 2003):

- advanced directional drilling systems and measurement systems that 'see' deeply into the formations during drilling
- intelligent down-hole sensors
- 4D seismics (time series of 3D seismics to investigate the effect of production on the reservoir)
- intelligent sensors for conditions and processes on the rig and on the seabed
- new data transmission technologies and optical cables connecting offshore and onshore locations

---

2   A device that turns the drill string. The top drive is suspended from the hook.
3   Adjusting the borehole position (inclination and azimuth angles) on the fly.

- the development of operations centres for operating and service companies, with advanced audiovisual systems and interfaces
- Internet-based ICT systems and international standards for electronic sharing of information that make possible the collection and exchange of real-time data between the offshore operations and the onshore locations.

This list of innovative practices enabled by IO indicates that it involves much more than the innovative use of ICT. Integrated Operations are also about sensor technologies, tools for interpretation, diagnosis and automation, and a new division of labour that implies a redistribution of skills and tasks in the organisation. Thus, IO do not only support the information infrastructure, but the whole system of *distributed cognition*.[4] Acknowledging the developments that have taken place within the industry during recent decades, however, it may be debated whether IO represents a paradigmatic shift or if it is merely a continuation of an ongoing change process.

In this book, the latter view is adopted. Further, IO is broadly conceptualised as a constellation of new technologies and work processes. More sophisticated definitions exist, but rather than framing the research by sophisticated, principal definitions, the strategy in this book is to explore a clearly defined *part* of IO and to investigate how this part of Integrated Operations may influence the safety of operations. The delimitation is made with respect both to the concrete sociotechnical innovations and the operational domain into which they are introduced.

Understanding IO as a historical phenomenon rather than a revolution reduces the risk of attributing special causal powers to IO and disregarding the significance of a historically developed constellation of technologies and work processes that forms a context for each new initiative for change. Accepting such a historic perspective, even the 15-year time span of OLF's (2005) IO projection seems extremely short:

> From a traditional operations regime with limited integration between the actors, the industry is believed to integrate the onshore and offshore work processes in what is labelled *Generation 1*. In this phase, spanning approximately ten years, self-sustainable fields, specialised onshore units and periodic onshore support will develop into integrated onshore and offshore centres and processes and continuous onshore support. Whereas Generation 1 is thought to facilitate integration across on- and offshore *within* companies, *Generation 2* is believed to overlap and continue Generation 1 to promote integration *across* companies through integrated operator and vendor centres, in addition to automated processes and digital services, where the operational concepts include onshore centres delivering a large portion of the services required to operate a field over the web on a 24/7 basis. (OLF, 2005)

---

4   See Hutchins (1995a, b), Giere (2002) and Giere and Mofatt (2003).

Although Figure 3.1 is informative and illustrates some important trends, the process of finding an optimal collaborative regime between all involved actors in such a heterogeneous innovative industry as the offshore industry is more likely to be a continuous process than one that will come to an end within the next three years.

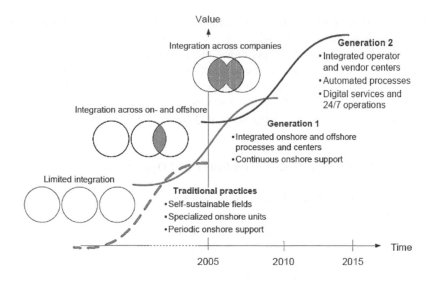

Figure 3.1   **Expected trajectory of IO implementation, as seen by OLF (2005). For an alternative way of characterising the IO trajectory, see Edwards et al. (2010)**

Source: Reprinted with kind permission from the Norwegian Oil and Gas Association.

3.3 Discourses of Integrated Operations

Although improved safety to some extent *defines* Integrated Operations by entering into some of the most canonical definitions, IO also lends its potential to many other phenomena and discourses: as a concept it cannot be seen in isolation from the power it lends to these discourses.[5] The *faster, better, safer* discourse is perhaps the one that is most strongly articulated by the industry itself. Here, IO represents a strategy to strengthen the industry in terms of effectiveness, safety and competitiveness through technological means. Few worries are indicated among the contributors to this discourse, perhaps the most central being the concern that *hesitation to adopt* the technologies may delay the change process.

---

5   Discourses of IO is a main topic in Vinge (2009).

Among those hesitators one may find those who indicate a discourse of scepticism that IO is a means of moving more functions and people from offshore to onshore, thus revoking many of the privileges associated with offshore work. However, while the relocation of functions is surely on the agenda,[6] the motives may of course be more than merely pecuniary.

The collaboration and communication discourse is perhaps strongest indicated by research and consultant communities. In this discourse themes like shared situation awareness, decision-making, collaboration in distributed teams and trust are central phenomena for investigation.

The concept of Integrated Operations, the analyses and the envisioned effects can be difficult to grasp not only because of the many different stakeholders and agendas. It can also be difficult to distinguish between empirical descriptions, analyses and visions. Identifying the discourses that IO initiatives belong to is a useful start when one is navigating in the IO literature and discussions.

3.4 Some Challenges of Research on Integrated Operations

The term Integrated Operations is often used in a broad and general manner, and this poses a problem when the effects of IO on safety are to be explored. In addition, lack of delimitation of the industrial domain is problematic because certain aspects of IO might influence different industrial domains differently. To make sure that the applicability of such explorations does not suffer from a too generic approach, three issues should be specified.

First, the theoretical and ontological orientation should be specified. This is necessary in order to evaluate the consistency of the research. Without accounting for this, it is possible to 'shop' between different theories and models that are associated with different and partly incompatible ontologies. In Chapter 9, the significance of consistency and accountability with respect to this issue is elaborated.

Second, rather than a general definition of Integrated Operations, a concrete part of the IO complex should be identified and studied. Since IO involves a wide variety of technologies and work processes, it is necessary to discriminate between its different aspects for an analysis to be accurate and meaningful.

Third, the petroleum industry involves a number of domains and professions. Treating them as one does not justify this heterogeneity and hence there is a risk that the findings will apply theoretically to a general representation of the industry, but not to the specific domains in a practical context.

Regarding the first issue, a sociotechnical orientation has successfully been adopted in several studies within the oil industry. Hepsø has explored the introduction of an integrated organisation and information technology concept in an oil and gas company and shown how the circulation of such a concept

---

6   See e.g. the abstract in Herbert et al. (2008).

depends on much more than the intrinsic qualities of the technology itself (Hepsø, 2002, 2009), He has also shown how the social nature of work involves a lot of articulation work which is often not addressed by standardised work processes that presuppose a seamless integration of data (Hepsø, 2006). Rolland et al. (2006) have studied how the integration of heterogeneous sources of information in the oil industry affects the conditions for collaboration, and how it may produce unintended consequences of disintegration. Almklov (2006, 2008) has studied how the production of knowledge of a subsurface oil reservoir involves a pragmatic, sociotechnical construction of objects whose existence and ontological status depend on the purpose they are capable of serving.[7]

Also, in this study the orientation of the research has been mainly along the lines of a sociotechnical discourse (see Chapter 6). This has affected not only *which* aspects of Integrated Operations have been studied, but also the *level* of description. For example, the elaboration of sociotechnical systems fluctuates between *descriptions of sociotechnical systems* and *prescriptions for how safety of sociotechnical systems can be obtained*, and some readers may be disappointed to find few clear answers to how a new technology or a new work process will affect the safety of those systems. As Latour (2005) has argued, in sociotechnical systems controversies and uncertainties dominate over facts and causal relations. Introducing a range of new sensors and tools for interpretation will definitely render the system more *complicated*, but how it will affect the *complexity* depends on the situated action that these tools will be a part of, and the degree to which they may be black boxed and contribute to stabilisation of the heterogeneous associations.[8]

In a commentary to Beck's (1992) notion of risk, Latour (2003a) elaborates on the concept of uncertainty. By comparing *risk* with *networks*, he illustrates the importance of tracing the heterogeneous networks; the safety of a sociotechnical system is always challenged by uncertainties of how different elements affect each other, and the stability[9] of the networks is not an intrinsic property of the system or its constituents, but a *result* of continuous sociotechnical work. Again, the elaboration fluctuates between descriptions of sociotechnical systems and safety of sociotechnical systems. This reminds us that safety and system description are

---

7   En passant, an interesting point is the similarity between the practical goal of IO, which is a 'tighter integration of technology, data, competency, activities and organisations' (RIO Project, 2010), and a theoretical point of Actor-Network Theory (ANT), namely that the relations between technology, data, competency, activities and organisations deserve more attention than they are being given in classical sociological approaches. Bearing in mind the assertion in Section 3.2 that IO is not a new paradigm, but a denotation for an ongoing sociotechnical development, it is tempting to interpret IO as a new theoretical description of the petroleum industry rather than a practical vision for the same industry.

8   For a more thorough description of the difference between complicatedness and complexity, see Section 2.1 and Chapter 8, Strum and Latour (1987) and Latour (2005).

9   The stability of sociotechnical systems is another way of expressing their predictability, which in turn is closely linked to safety.

not only closely related, but also a result of the work of the actors working *in* the system as well as e.g. the researchers working *on* it. The advice that is sought after – how can we make this system safe or what effect will this change have on the system's safety – may be replaced by a question: *How will the change process in question alter our descriptions of the sociotechnical system?* This is also the question that this book deals with. It may seem less ambitious at first sight, but I claim it is not. A good description is an analysis in itself.

Regarding the second issue, the part of Integrated Operations that is addressed by this book is constituted by tools for monitoring, interpretation and automation of the drilling process. This choice naturally also influences the third issue, namely the industrial domain, which in this study is the drilling operations. It leaves out such domains as maintenance, reservoir management and production optimisation.[10] The delimitations implied by issues two and three should not be seen as absolute, but as a point of departure for the study. As explicitly demonstrated in Chapter 7, such delimitations of domains and tools will in practice have to be crossed for a phenomenon to be adequately explained. Thus, these delimitations should be seen as an identification of the study object and a starting point from where one cannot decide in advance where to proceed; this is exactly what fieldwork following ANT[11] tradition is about.

---

10   As accounted for in Chapter 1, the choices of issues two and three were actually taken in the reverse order.

11   See Section 5.5.

Chapter 4
Drilling for Oil and Gas[1]

4.1 Two Central Actors – Petroleum Technology and Drilling and Well

Petroleum technology and *drilling and well* are two central professional communities in oil and gas companies that participate in different phases of a well construction process. The expertise of the two groups will be briefly reviewed here, while the groups' roles in the well construction process will be elaborated on in Section 4.2.

4.1.1 Petroleum Technology

The petroleum technology department is staffed by geophysicists, geologists, petrophysicists, reservoir engineers and production engineers. The *geophysicists* are central in interpreting the seismic data, and with the additional use of production data they refine the interpretations of the seismic data and produce cross-sections of the formations with stratification and faults. While the geophysicists are mostly concerned with structures, the *geologists* are concerned with the geological properties of the rocks, their history of deposition, their drillability, their likelihood of containing petroleum, and the creation or refinement of the geo-model. The *petrophysicists* are concerned with the reservoir's rock properties, such as porosity, saturation, permeability and electrical conductivity. These properties are explored by the use of, e.g., different well logs, core samples and seismic measurements. The *reservoir engineers* are interested in the volumes and composition of hydrocarbons in the reservoir, the flow conditions and the drainage strategies to extract as much hydrocarbons as possible from the reservoir. This also involves advising a well target and an appropriate well path through the reservoir. The *production engineers* are, as their title indicates, focused on the production of hydrocarbons from the reservoir. Central topics are thus the flow regimes from the reservoir into the well and from the well to the surface, in addition to the processing of oil, gas and water on the surface. They monitor the pressures in

---

1 This chapter is not meant to present petroleum technology and drilling in the fashion of an educational book. It is meant to give a brief introduction to some of the phenomena that characterise drilling operations, with the purpose of providing the reader with useful connotations for the subsequent reading. Some readers may perhaps struggle to see the use of some of the details, but I would urge them not to skip them; one cannot understand IO without understanding the field in which it is introduced. A more thorough elaboration of the topics of this chapter may be found in Hyne (2001).

the reservoir and plan and administer the production of different wells in order to optimise overall production from the reservoir. The production engineers work closely together with the reservoir engineers and well engineers from the drilling and well department.

4.1.2 Drilling and Well

The drilling and well department is responsible for the detailed planning and execution of the well construction, and the main staff consists of drilling engineers and well engineers. When we speak of onshore rig teams in this book, we refer primarily to these engineers and their supporting staff. The division of work in the department is connected to different phases of the drilling operations. The drilling engineers are responsible for the actual drilling of the well according to the individual drilling programme. This involves close cooperation with the operations geologist, the drilling contractor and the service companies. The well engineers are responsible for completing wells that have reached their targets and preparing them for production, and for well interventions in wells that are already in production. The well engineers work close together with the offshore rig crew, production engineers and reservoir engineers, in addition to various service companies.

4.2 The Well Construction Process – Conceptual Workflow

As an operating regime, Integrated Operations are expected to pervade the whole field of the upstream[2] oil and gas industry. What is often simply referred to as 'oil and gas exploration and production' represents an extremely heterogeneous domain with respect to the types of work and the professions involved. Before going into the details of the part of IO that is explored in this study, a brief review of the chain of work processes along the axis of upstream activities from exploration to production will provide a useful context for the reader. Although the description below is collected from one specific company and the details may vary across companies, it points to some key phases and collaborative practices in the process that have a more generic validity.

The first step in the well construction process is the seismic mapping of the overburden formations and the reservoir containing the hydrocarbons. Geophones and receivers attached to wires are drawn by ships over the areas that are mapped. The geophones produce sound waves that travel through the sea and through the rock formations, where borders between different rock types produce patterns of reflections that are registered by the receivers and thereafter projected onto paper and into computer models. The resulting two- and three-dimensional 'images' of the formations serve as starting points for the planning of the wells.

---

2   The exploration and production part of the industry.

The rig team first gets involved with the operations when it is decided that a new well will be drilled and the start-up meeting is arranged. This marks the start of the *well planning phase*, the first phase of the well construction process shown in Table 4.1.

The first, conceptual part of the well planning phase is run by a well planning team composed of experts covering the different phases throughout the lifespan of the well: geophysicist, reservoir engineer, petrophysicist, evaluation geologist, programme well engineer, operations geologist, programme drilling engineer, production engineer, service contractor and drilling contractor. In addition, the well planning team can seek support from other resource persons and communities, such as a subsurface support centre. The well project coordinator in the conceptual phase is a representative from the petroleum technology department.

The result of the work in the conceptual phase is the commission of a specific well. The specifications include a well target, a well path and a methods selection. In the next part of the well planning phase, the *detailed planning and engineering phase*, coordination responsibility is handed over to a representative from the drilling and well department, who starts the detailed planning of the drilling operations.

Table 4.1   The well construction phases

1. Well planning phase		2. Well operations phase	3. Well evaluation phase
Conceptual phase	Detailed planning and engineering phase	Execution phase	Conclusion phase

The main result of the detailed planning and engineering phase is the *individual drilling programme*, which is a detailed description of the well construction process; the division of the well into *sections*,[3] the composition of the *bottomhole assembly*,[4] the sensors and the *drilling fluid*[5] characteristics for each stage of the construction process is described in the drilling programme. When the drilling programme is formally approved, a pre-operations meeting indicates the transition

---

3   This is explained in Section 4.3.

4   The lower portion of the drill string consisting of the bit, a mud motor, stabilisers, drill collars and directional drilling and measuring devices.

5   The drilling fluid is pumped down through the drill string and returns up in the annulus – the space between the drill string and the well bore. It serves the purpose of transporting drilled cuttings – pieces of rock – from the well and preventing fluids (oil, gas or water) from flowing from the subsurface formations and into the well (Hyne, 2001). The drilling fluid is custom-made for each well.

into the *well operations phase*, where the plans are executed and the well is constructed. When the operations enter this phase, the main responsibility is handed over from the well planning team to the well operations team.

The well operations team consists of the rig team and the operations geologist, in addition to the drilling contractor and the service companies. The rig team is staffed mainly by drilling engineers. In collaboration with the offshore rig crew, which is executing the drilling programme, and with support from the geologists and the reservoir engineers, the drilling programme is executed and transformed into a well that can be completed and prepared for the next phase of oil and/or gas production, water injection to support other producing wells, or other uses that the specific well is designed for.

After completing the well, the programme drilling engineer is responsible for filing the *detailed operations procedures* and a compiled *daily drilling report*, and the geologist will produce a *final geological and reservoir technical end of well report* including a geological, geophysical and petrophysical evaluation of the well data. At this stage, the drilling and well department has finished its job and can hand the well over to the *production department*.

4.3 The Well Operations Phase – Drilling in Practice

In the well operations phase, the drilling programme is executed and the physical well is constructed. The work in the operations phase is basically divided between two communities. The onshore rig team has been central in the design phase of the well construction process, and is also central in the execution phase. It is the offshore rig crew, however, that carries out the drilling programme in practice, by executing the detailed operations procedures prepared by the rig team. The division of work and the collaboration between these two communities will be dealt with in Section 4.5, after a review of the main activities involved in the drilling operations and the main risks that are associated with them.

4.3.1 Construction

Reaching from the sea bed to the target somewhere in the reservoir, the measured depth of a well can exceed 10,000 metres. A well may be vertical or it may be deviated, reaching several thousand metres horizontally. It may penetrate many different zones involving many different challenges. One challenge is the geology. While some rocks, such as sandstone, can be drilled fast and with little wear and tear on the drill bit, limestones, for example, can be hard to penetrate and may wear out the bit quickly. Another challenge of a more structural kind is faults that may make geological interpretation difficult, because expected layers may appear twice or not at all due to local displacement. Fault zones are also potential loss zones, where drilling fluid can escape into cracks in the formations and threaten the hydraulic balance in the well. A third challenge is that different formation zones may contain gases or fluids at different pressures. This makes it difficult to maintain the primary well control by conventional drilling methods using a homogenous mud column. In zones where the formation pressure is higher than the pressure of the drilling fluid, formation fluids can flow into the well, whereas drilling fluid can be lost to formation zones of lower pressure. These and other challenges are part of the context of the construction process, and represent the uncertainties that make the drilling programme a resource rather than a dictate (Suchman, 2007).

The well is constructed in sections. The first section is drilled with a drill bit of a large radius, typically 36" (inches). Thereafter, holes of smaller and smaller diameters will be drilled until the target is reached by the end of the section with the smallest hole diameter. A sequence of hole diameters may be 36", 26", 17½", 12¼", and 8½", but some of these sections may be skipped to make the operations faster. Before a new section is drilled, the last section is secured with a *casing* so that the hole will not collapse and the drilling fluids are safely contained within the annulus between the casing and the drill string. The corresponding diameters of these casings will be 30", 20", $13^3/_8$", and $9^7/_8$", as illustrated in Figure 4.1.

The casings are attached to the hole wall with cement that is pumped down through the casing and up into the outside of the casing's lower part.

4.3.2 Steering

In addition to the sequential aspect of the construction, the spatial placement of the well is decisive for the success of the operation. First, the well must follow a path that does not collide with other wells in the area. Second, optimal production of hydrocarbons presupposes an optimal placement of the well in the reservoir. Conditions that make both steering and the accurate measurements of positions especially challenging are the considerable lengths and deviances of wells. In addition, the drilling of wells in areas that are not much explored, as is the case with *wildcat*

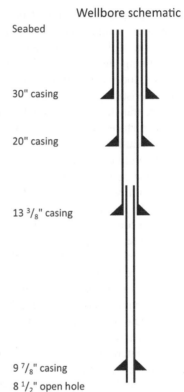

Wellbore schematic

Seabed

30" casing

20" casing

$13^3/_8$" casing

$9^7/_8$" casing
$8^1/_2$" open hole

Figure 4.1 Well schematics. The figure shows casing diameters. In the 8½" open hole section there is no casing. Figure from drilling programme

drilling, may involve a considerable geological and structural uncertainty. In the last section of a production well, the reservoir section, all these challenges may coincide when the drilling operation involves *geo-steering*. During geo-steering, the operational geologist (offshore) is central as the positioning in the reservoir is determined by analysing the geological properties while drilling. These properties can be collected from return cuttings[6] and from the measurements while drilling (MWD) and logging while drilling (LWD)[7] tools.

4.3.3 Interpretation, Testing and Diagnosing

The feedback of drilling data during the operations forms the basis for the continuous evaluation of the drilling process. These drilling data constitute the only accessible representation of the down-hole conditions, and monitoring and interpreting these data is thus a key process in the operations. In down-hole sensors, located in the bottom-hole assembly,[8] geological properties and process characteristics (e.g. mud properties, bottom-hole pressure, torque/drag, weight on bit) are transformed into mud pulses or electrical signals, which are transported to the surface by the circulating mud or through wired drill pipe. Some of the data is real-time data, some is lagged. Some data is of high resolution, some of a low resolution. Some data is aggregated, some reflects one single property of one identified entity (Almklov et al., 2012). The work of interpreting such data-sets involves a recombination of fragmented, heterogeneous data into constellations that make sense. That recombinations *make* sense underscores the point that they do not reconstruct a reality in the offices identical to *the* reality deep down in the well; sense has to be *created* through re-presenting the well and the drilling process. The recombination is made with a purpose that is not present in the subsurface domain from which it is collected. It could be said that the recombination is a purposive representation of the well and the drilling process. Consequently, the recombination may be called an interpretation, in contrast to a blueprint that would not contain any element of interpretation and would therefore be redundant and not serve any purpose.

The interpretation work is a core component of the drilling engineers' work; to make sense of the drilling process and the formations that are drilled through in order to safely and effectively construct the well. The interpretations form the basis for diagnosing situations where the drilling data do not correspond to expectations. Such lack of correspondence may be indicated during ordinary operations or by the many tests that are undertaken to calibrate the simulations and interpretations.

---

6   Drilled rock chips that are transported to the surface by the circulating drilling fluid.

7   'Measurements-while-drilling measures well properties such as azimuth and deviation and logging-while-drilling measures rock and fluid properties' (Hyne, 2001: 329).

8   Not all data come from the sensors in the bottom-hole assembly. Some sensors are also located topside, and some data are extracted from samples of returning drilling fluid and cuttings on the rig.

One example of such tests could be the *leak off test*, which is a method to test the strength of a formation, i.e. what pressure it can withstand from the drilling fluids before it fractures.

4.3.3 Work Processes

The drilling operations are structured by daily meetings where the onshore and offshore organisations update each other on the status of the work, and review and adjust the plans for the upcoming work.[9] In addition, detailed operations procedures that describe how the tasks in the drilling programme are to be carried out are quality checked in separate meetings. Every morning, the onshore rig team is updated on the operations the last 24 hours in the form of a daily drilling report (DBR[10]) where key parameters characterising the operations, its progress and any problems are listed. For the onshore drilling engineers, glancing at the DBR when arriving at work in the morning is a way of getting a quick overview over the activities and contingencies since they left work the day before.

In addition to the regular and irregular meetings, the rig team monitors the operations in real-time, with real-time data. Although the drilling engineers do not usually monitor the operations from minute to minute or hour to hour, the presence of the information either on each engineer's personal computer or on larger, shared screens makes it possible to stay updated with only a quick glance at the screen. This access to drilling data also brings the operations closer to the onshore office. Although the data have travelled far, the real-time access reduces the difference in the way the offshore rig crew and the onshore rig team experience and interpret the operations.

The offshore rig crew and the onshore rig team, in addition to the service providers, are the central actors in the drilling operations, but other actors will normally also be involved in consultation during the operations. A growing number of support centres have received an increased focus in later years. One example of such centres is what is called a *Subsurface Support Centre* (Løwén et al., 2009). This centre offers support to the onshore rig team at their request.[11] The expert centre may give advice in the planning phase of a well project, for example, or assist in solving specific problems that may occur during drilling. In addition to the companies' own expert centres, service companies are deeply involved in the operations in connection with issues such as choice of bits and drilling fluid, and more complicated issues such as drilling optimisation.

---

9   The morning meeting is primarily for reporting; discussions and detailed planning is normally done in separate meetings. The need for such separate meetings may be identified in the morning meetings, however.

10   Norwegian: *'daglig borerapport'*.

11   The centre is only fully staffed during daytime.

4.4 Drilling Operations Risks

Throughout the lifespan of a well, different phases are associated with different risks. Drilling operations are associated with an array of well-known problems that may threaten the success of the well construction. Initially, some of these problems represent merely a delay in terms of time and money, while other problems are more acute, with the exposure of immediate risk. However, most drilling problems involve the occurrence of an unplanned or unforeseen event. Common for these events, without respect to their immediate risk potential, is that they challenge the operational control. The drilling engineers must deviate from the original drilling programme in order to cope with the problem and regain control. Such contingencies change the premises on which the drilling programme was originally based, and they introduce new uncertainties to the operations.

Another point is that the most serious risk in drilling operations, blowouts,[12] do not necessarily appear all of a sudden, but might develop from less serious problems. Thus, the prevention of blowouts also involves a controlled handling of less serious problems. In the end, avoiding serious accidents is about maintaining control of the operations throughout their different phases. In this study, maintaining control refers to the ability to cognitively understand, at any time, what is going on and what needs to be done to bring the situation, or the relevant parameters, back within the safe operational window. Below, some typical drilling problems (Hyne, 2001) that challenge the control of the operations are briefly reviewed.

4.4.1 Shallow Gas

Small pockets of gas in the upper formations, far above the reservoir, can represent a serious risk if they are not identified in advance and met with adequate measures. One consequence can be that the rig floor gets flooded with drilling fluid. More serious consequences occur with ignition of the gas. Generally the overburden formations are not mapped in the same detail as the reservoir formations, and the presence of shallow gas may therefore be unexpected.

4.4.2 Corrosive and Poisonous Gases – CO_2/H_2S

Corrosive gases such as CO_2 and H_2S may be released and flow into the well during drilling. These gases may react with the drill string and weaken it by a process of hydrogen sulphide embrittlement. Detection of such gases, and countermeasures in case of their presence, is necessary to prevent damage to the drill string. The gases are also poisonous, and their transport to the drill floor may represent a danger to the rig crew.

---

12   'An uncontrolled flow of fluids up the well' (Hyne, 2001: 280).

4.4.3 Sloughing Shale and Stuck Pipe

Some shale types react with water from the drilling fluid and swell into an impermeable pulp that is not easily removed from the well and that may cause the drill string to get stuck in the well (differential sticking[13]). The prevention of differential sticking is connected to aspects of well control. One method of prevention is to drill with sufficient low mud weight to reduce the overbalance pressure.[14] This is not completely unproblematic since overbalanced drilling is a central method of controlling the formation's fluid pressure to prevent kicks.[15]

4.4.4 Lost Circulation

When drilling into a porous or highly fractured formation, drilling fluid may be lost to the formation. As a result, the vertical height of the fluid column is reduced, leading to an influx of formation fluids from other zones. In the worst case, a situation of lost circulation may turn into a loss of well control and an uncontrolled blowout.

4.4.5 Blowout

In the well construction phase, uncontrolled blowouts represent the risk with the most serious potential. A blowout endangers the lives of the personnel, the environment and great material values, in addition to the possible loss of the well and all future income from it. A blowout occurs when the drilling fluid pressure is not sufficient to contain the formation fluids. The initial stage of a blowout is a kick, where formation fluids – oil, gas or water – flow into the well. If the kick gets out of control and the formation fluids flow freely to the surface, the situation is called a blowout.

About 50 per cent of blowouts occur during tripping out.[16] When the drill string is lifted out of the hole, the level of drilling fluid in the well, and thus the pressure exerted on the bottom, is reduced. This can cause formation fluids to enter the well. Another mechanism that may lead to a kick and a blowout during tripping out is when the formation fluids are sucked into the well due to a swabbing effect.

---

13 'A condition whereby the drill string cannot be moved (rotated or reciprocated) along the axis of the well bore. Differential sticking typically occurs when high-contact forces caused by low reservoir pressures, high well bore pressures, or both, are exerted over a sufficiently large area of the drill string' (Schlumberger online oilfield glossary, www. glossary.oilfield.slb.com)

14 The drilling mode is overbalanced when the pressure of the drilling fluid is higher than the fluid pressure in the formations.

15 A kick is the flow of fluids from the subsurface rocks into the well (Hyne, 2001).

16 Pulling the drill string out of the well.

4.4.6 Levelling the Risk Hierarchy

Although a blowout is definitely the risk with the most damaging potential in drilling, a classification of risks based on typology and damaging potential is far from unproblematic. One of the companies in the study distinguishes between risks connected to HSE, economy (time and cost) and objective (goal achievement). However, the causation of problems can be very complicated, and different problems can have common causes. Also, a minor problem can be the first step in the development of a more serious problem. A stuck pipe, for example, which is basically a risk with reference only to non-productive time and economic losses, can under certain conditions lead to the development of a blowout.[17]

By categorising risks into HSE risks and risks of reduced effectiveness, the connection between HSE and effectiveness is concealed. The trade-off nature of this connection, which is addressed in Hollnagel's (2009) ETTO model, implies that the countermeasures in a situation where the operations are lagging behind are likely to influence the balance between efficiency and thoroughness and thus indirectly influence safety. In operations that lag behind the time schedule, and where the operators and decision-makers feel the pressure from other stakeholders to deliver good results, choices may be made that increase the HSE risk. Vaughan (1996) exposed this mechanism in her review of the Challenger Launch Decision, and also in the Deepwater Horizon accident that occurred in the Gulf of Mexico on 20 April 2010 the operations were behind schedule and over budget, something that may have contributed to shortcuts being taken (Graham et al., 2011).

Consequently, in this book the risks that are considered are not only HSE risks with an obvious and immediate damaging potential. Work processes that involve risks connected to objective and economy, or non-productive time in general, are considered relevant.

4.5 Different Disciplines and Different Perspectives

One characteristic of drilling operations is the large number of actors involved. The number of specialists and consequently the division of labour in the operations has increased steadily in parallel with technological development. That there are few generalists left in the industry was mentioned by many of the informants, and although the specialisation and division of labour is highly necessary to accomplish complicated work tasks, it also creates challenges of meshing together these tasks and preventing the interfaces between them from affecting the work negatively.

The interfaces between discrete work tasks are seldom simple and unambiguous. An example of this is the work of choosing an optimal drill bit for

---

17   A case of this could occur in a situation when the blowout preventer needs to be engaged. If the drill string is stuck in such a position that the shear ram cannot cut it, a serious situation may develop.

a section. Whereas the bits were fewer and the geological analyses less accurate in earlier days, today a range of different drill bits with different properties exists and the geological prognoses based on a combination of seismic mapping and data from other wells in the same area produce models with higher resolution than before. Deciding on an optimal drill bit is, nevertheless, not a straightforward task. Whereas the drilling engineer must focus on the time spent on the whole bit run[18] and keeping that as low as possible,[19] the bit vendor will have a different set of success criteria. Actual discussions between a bit vendor and a drilling engineer that occurred during the fieldwork illustrate this point: a bit vendor eagerly recommended a new type of bit that had not been tried before. One argument for this was that the bit they had used for similar geological conditions in a previous and nearby well had worked unsatisfactorily, and that this new bit was developed specially to drill fast under similar conditions. The drilling engineer's response was that the rig team organisation felt that it was necessary to have a good result for this well because the record for the last wells was not impressive, and therefore they wanted to 'bet on a safe horse'; a bit that was not necessarily very aggressive but rather one that would perform acceptably under various conditions. Since the bit recommended by the vendor was designed to be especially aggressive under the conditions that were expected, but could potentially wear out and have to be changed if they ran into unexpected hard stringers of limestone, it was perceived as too risky by the drilling engineer. The bit vendor then mobilised more arguments supporting his view: the development of new and more effective bits is impossible without participation from operating companies through testing of the bits under realistic conditions, and he felt that the operating company had not been sufficiently committed during the last months. Without expressing it clearly, he indicated that those who do not contribute cannot expect to get access to any successful results, at least not without paying a high price. The discussion between the two professionals from different disciplines clearly illustrates how the optimal meshing of divided labour is complicated.

In drilling operations, drilling and well and petroleum technology are two central disciplines whose work methods can sometimes be difficult to mesh together. The disciplines are organised into separate departments and their perspectives are often in conflict with each other. Drilling and well's and the drilling engineers' job finishes when the well is completed, and the success of their job is mainly measured in the time spent on the well project. The petroleum technology department, on the other hand, with its reservoir engineers and its geologists, is not measured by how long time it takes to drill the well. Their success is measured in how much oil and gas the well produces during its lifespan, which depends highly on the well's placement in the reservoir.

---

18 A bit run is the drilling of a part of the well undertaken with the same bit, from running in to pulling out.

19 The day rate for renting a rig can be several hundred thousand dollars, and the cost of operations is thus largely decided by how long the operation takes.

Thus, whereas the drilling department has a short time horizon and relatively low accuracy requirements for the well trajectory, the petroleum technology department's requirements for accuracy of the well paths are more demanding, and the time horizon over which the success will be evaluated is longer. Still, both disciplines' contributions to the drilling operations are inextricable linked to each other, and the accomplishment of this work depends on their collaborative work. Collaboration under such conditions is difficult to standardise and is in the elaboration of the empirical data in this book described as *articulation work* (Strauss, 1985).

Chapter 5

Safety of Sociotechnical Systems and Sociotechnical Work

Viewing Integrated Operations as sociotechnical systems, how may we analyse our study object further? We might find many valuable insights in the historical and contemporary research on industrial and organisational safety, but to fully understand the dynamics of integrated offshore operations we will need to supplement these with a methodological tool that accounts for the more fundamental ontological characteristics of sociotechnical systems. In this chapter and the next we will equip our toolbox with safety theories and methodological perspectives, so that we will eventually be in a position to analyse the effect of new technologies and work processes on the safety of the operations in general, and to apply this insight in our analyses of the cases presented in Chapter 2.

5.1 Accident Genealogies

A review of the search for causes of accidents during the last 50 years shows a pronounced trend (see Figure 5.1). Whereas *technology and equipment* constituted the dominant cause in the 1960s and *organisation* was hardly represented at all, in the mid-1990s *human performance* had taken over as the dominant cause by far, while technology and equipment had dropped significantly as an attributed cause, and *organisation* continued a slow but evident increase. At the time of writing, a combination of trend analysis and qualified speculation may indicate a future trend; the attribution of failure to human performance and technology and equipment is declining, while the attribution of failure to organisation continues to increase (Hollnagel, 2004).

Since what you find often corresponds to what you look for, it is not surprising that the changing dominance of causes to some degree corresponds to the contemporary dominant safety theory discourses.

Three main types of accident models are identified by Hollnagel (2004): the *sequential* accident model, the *epidemiological* accident model and the *systemic* accident model. The sequential model assumes that accidents result from a simple linear chain of events or factors, each event triggering the next one. One example of a sequential accident model is Heinrich's (1931) Domino theory, illustrated in Figure 5.2.

% Attributed cause

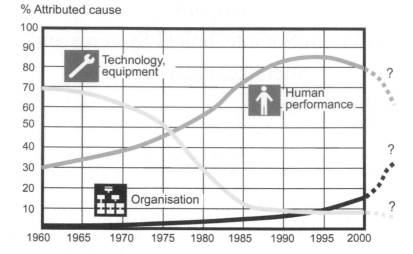

Figure 5.1   Trends in attributed accident causes (Hollnagel, 2004)

Source: Reprinted by permission of the Publishers, in *Barriers and Accident Prevention* by Erik Hollnagel (Farnham: Ashgate, 2004), p. 46, Copyright © 2004.

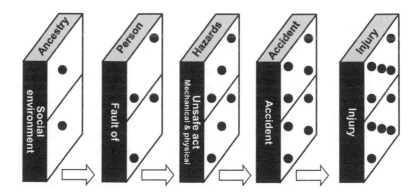

Figure 5.2   The Domino theory (after Heinrich, 1931)

Source: Reprinted by permission of the Publishers, in *Barriers and Accident Prevention* by Erik Hollnagel (Farnham: Ashgate, 2004), p. 49, Copyright © 2004.

In the sequential accident model the root cause is looked for by searching backwards in the causal chain until the 'first domino' is identified.

While the sequential model assumes a simple linear chain of events, the chain of events in the epidemiological accident model assumes a complex linear relationship (Hollnagel, 2004). Here, accidents are seen as the outcome of a combination of latent and manifest conditions, analogous to the spreading of diseases. The epidemiological model differs from the sequential model in four ways. First, the notion of human error is replaced with *performance deviations*. This implies that errors can occur in both technological components and humans. Also, the term *deviation* indicates a less categorical view on the rationale for actions than the term *human error*. Second, the introduction of *environmental conditions* nuances the view on root causes. Third, *barriers* are introduced as a means of stopping the development of the accident at any stage in the process. Fourth, the introduction of latent conditions that are present in the system before any accident sequence sets off is recognised. These conditions do not cause any harm to the system under normal circumstances, but when an accident starts to develop they contribute to the accident development. Figure 5.3 illustrates the epidemiological accident model.[1]

The third main type of accident model in Hollnagel's (2004) classification scheme is the *systemic accident model*. This focuses on the performance of the system as a whole without attributing linear causes/effects to the explanations of accidents, addressing non-linear, non-decomposable systems. Accidents are thus not understood as slowly developing in a sequential manner, but as emergent phenomena resulting from unexpected combinations of conditions that in themselves are not seen as abnormal or unusual. Compared to the epidemiological model, the arrow indicating the direction of causality has been removed. This

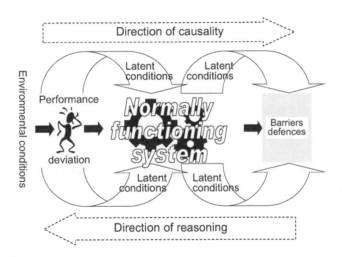

Figure 5.3   The epidemiological accident model

Source: Reprinted by permission of the Publishers, in *Barriers and Accident Prevention* by Erik Hollnagel (Farnham: Ashgate, 2004), p. 55, Copyright © 2004.

---

1   Reason's (1997) Swiss cheese model is often used as an example of the epidemiological accident model. It is briefly discussed in the next chapter, where it is used in a different classification system.

reflects the assumption that there is not one unambiguous sequence of action in the system; each event may be both preceded and followed by several events (see Figure 5.4). Another difference is the introduction of the blunt end/sharp end dichotomy. While the sharp end refers to those who directly interact with the work processes in question and perform the tasks, the blunt end involves those who affect safety by influencing the constraints and resources at the sharp end. What is the sharp end and what the blunt end is relative to what action is observed, but a generic list of actors and factors listed from the sharp to the blunt end in a certain work process may include operator, local workplace factors, management, company, regulator, government and morals and social norms (Hollnagel, 2004: 63). Figure 5.4 is a simple and generic illustration of the systemic accident model.

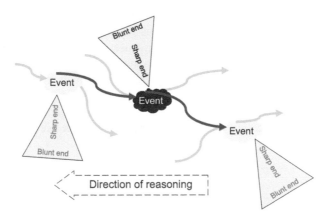

Figure 5.4   The systemic accident model

Source: Reprinted by permission of the Publishers, in *Barriers and Accident Prevention* by Erik Hollnagel (Farnham: Ashgate, 2004), p. 60, Copyright © 2004.

5.2 Generic Models of the Genealogy and Handling of Organisational Accidents

Rosness et al. (2010) apply a different classification in their treatment of organisational accidents. Their review of the different understandings of accidents and safety focuses on generic models of the genealogy of accidents and how they are handled. Six distinctive perspectives in the organisational accidents' discourse are identified.

The energy and barrier perspective (Gibson ,1961; Haddon, 1980) calls attention to the physical energy involved in accidents, and the role of barriers in preventing them. According to this perspective, accidents occur when 'objects are affected by harmful energy in the absence of effective barriers between the energy source and the object' (Rosness et al., 2010: 113). Three different strategies for loss reduction are identified: reduction of the hazard, strengthening of the barriers and protection and rehabilitation of the victims. Defence in depth is achieved by applying several barriers directed towards different parts or processes of the system. However, as Reason's (1997) Swiss cheese model illustrates, even such barrier systems may fail.[2]

Barriers can be defined as 'physical and procedural measures to direct energy in wanted channels and control unwanted releases' (Johnson, 1980: 508), including procedures, work permit systems and other administrative measures (Rosness et al., 2010). In this model, failures are classified as *active failures* and *latent conditions*. Active failures are 'unsafe acts (errors and violations) committed by those at the "sharp end" of the system ... It is the people at the human-system interface whose actions can, and sometimes do, have immediate adverse consequences' (Reason, 1995: 82).

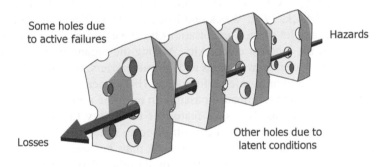

Successive layers of defences, barriers, and safeguards

Figure 5.5   An accident trajectory passing through corresponding holes in the layers of defences, barriers and safeguards

Source: Reprinted by permission of the Publishers, in *The Human Contribution* by James Reason (Farnham: Ashgate, 2008), p. 102, Copyright © 2008.

Latent conditions, on the other hand, are

> created as the result of decisions, taken at the higher echelons of an organisation. Their damaging consequences may lie dormant for a long time, only becoming evident when they combine with local triggering factors ... to breach the system's defences. (Reason, 1995: 82)

---

2   Note that the Swiss cheese model is not originally developed within the energy barrier perspective, but within an organisational or systemic perspective.

The *Normal Accident Theory* (NAT) (Perrow, 1999) is a framework for understanding how specific system characteristics affect an organisation's proneness to system accidents. According to Perrow (1999), the control of systems with tight couplings depends on a centralised control regime, whereas the control of systems of high interactive complexity demands a decentralised control regime. These demands thus become conflicting and insurmountable when a system is both tightly coupled and highly complex.

The combination of system characteristics that affects the ability to successfully control a system is illustrated in Figure 5.6; the industries placed in the upper, right-hand corner, such as nuclear plants, are so tightly coupled and interactively complex that large accidents are, according to Perrow, impossible to avoid.

Whereas Perrow's recommendation is that due to their intractability, tightly coupled, interactively complex systems should simply be avoided, the *High Reliability Organisation* (HRO) approach (La Porte, 1996; La Porte and Consolini, 1991; Weick, 1987; Weick and Roberts, 1993; Weick and Sutcliffe, 2007) points to examples of such organisations having a remarkably good safety record in spite of the inherent complexity. Among the organisations that have been identified are nuclear power plants and aircraft carriers. Instead of focusing on characteristics that inhibit safe performance, HRO researchers focus on the features that render safe performance possible in even the most tightly coupled and complex organisations. One such feature is the ability to reconfigure the organisation in times of crisis from a centralised to a decentralised structure to mobilise the relevant expertise, wherever it may be localised within the organisational structure. Another feature is organisational redundancy, which is achieved by crew members with overlapping tasks and competences who

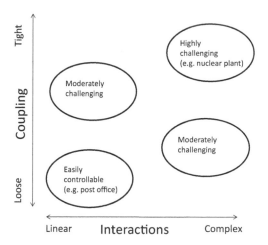

Figure 5.6   Couplings and interactions in different industries

seek advice from each other and intervene to recover each other's erroneous actions (Rosness et al., 2010). Another way to describe HROs is to refer to five characteristics: preoccupation with failure, reluctance to simplify interpretations, sensitivity to operations, commitment to resilience and deference to expertise (Weick and Sutcliffe, 2001).

A fourth perspective is that accidents are caused by a breakdown in the flow of information, a perspective that in Rosness' review is represented by Turner's (1978) *theory of man-made disasters*. In this perspective, sometimes called the disaster incubation theory, the focus is on the way disasters *develop* rather than simply occur, through a long chain of events that leads back to root causes of misperception among individuals and lack of information flow. The chain of events unfold during the *incubation period*, which is allowed to progress unnoticed due to 'rigidities in institutional beliefs, distracting decoy phenomena, neglect of outside complaints, multiple information handling difficulties, exacerbation of the hazards by strangers, failure to comply with regulations, and a tendency to minimize emergent danger' (Turner, 1976: 378).

Unnoticed failures may travel and magnify as *anti-tasks*[3] through the organisational decision hierarchy or power hierarchy, thus using the administrative and operational structures and procedures meant for normal and safe work processes in an unintended way.

In the *decision-making perspective*, the focus is on handling conflicting objectives. Under the pressure of conflicting objectives such as acceptable workload, financially acceptable behaviour and functionally acceptable behaviour with regard to risk, the performance of organisations and their groups and individuals may drift. Rasmussen's (1997) drift model (see Figure 5.7) illustrates how the boundary of safe performance and thus the error margin is challenged by other objectives.

A sixth perspective on accidents and safety is *Resilience Engineering* (Hollnagel, 2009; Hollnagel et al., 2008, 2006). A central credo in this perspective is that the efficiency–thoroughness trade-off (ETTO) (Hollnagel, 2009) is an integral part of all work, and that a successful outcome depends on this balance. The genealogy of failures is not different to that of successes; 'both failures and successes are the outcome of normal performance variability' (Hollnagel et al., 2008: xii), and the challenge is to 'find ways to reinforce the variability that leads to successes as well as dampen the variability that leads to adverse outcomes' (Hollnagel et al., 2008: xii). Responding to this challenge requires four essential capabilities, also called the four cornerstones of resilience engineering: knowing what to *do* – the ability to address the *actual*. Knowing what to *look for* – the ability to address the *critical*. Knowing what to *expect* – the ability to address the *potential*. Knowing what has *happened* – the ability to address the *factual*. The main characteristics of resilience engineering appear in Figure 5.8.

---

3   Other terms used by Turner (1978) are 'orderly errors' and 'negenthropic mode'.

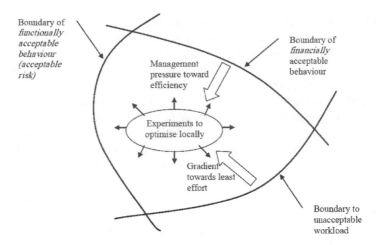

Figure 5.7   Under the presence of strong gradients behaviour will very likely migrate toward the boundary of acceptable performance (Rosness et al., 2010, adapted from Rasmussen, 1996)

Source: Reprinted with permission from Rosness/SINTEF.

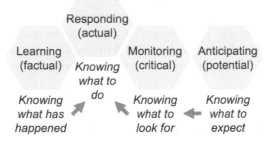

Figure 5.8   The four cornerstones of resilience

Source: Reprinted by permission of the Publishers, in *Resilience Engineering in Practice* eds Erik Hollnagel, Jean Pariès, David D. Woods and John Wreathall (Farnham: Ashgate, 2011), p. xxxvii, Copyright © 2011.

The assertion of the common genealogy of failures and successes has important implications for studies of how accidents can be prevented or handled; since there are far more successful operations than failures, there is a much better basis for learning by studying successful operations than failures. Thus the advice to study and explain successful operations is a recurring feature of resilience engineering.

5.3 A Comment on Conflicting Models of Safety and Accidents

Much safety theory literature involves models of how accidents occur and how they can be prevented, as the reviews in Sections 5.1 and 5.2 demonstrated. These models and prescriptions are not unproblematic. Many of them are incompatible. Some examples: NAT and the HRO approach come to different conclusions with respect to industrial processes with tight couplings and high interactive complexity; the simple linear model and the systemic model describe very different system characteristics and accident genealogies; resilience engineering treats interactive complexity as a relative (epistemological) entity, whereas NAT refers to an objective (ontological) description. Furthermore, there is a tendency to base the choice of model on the theoretical orientation of the researcher rather than to adapt the research to the characteristics of the industrial process being studied. For instance, Leveson et al. claim that most cases studied by HRO scientists are erroneously categorised: 'using engineering definitions, the design of most of the engineered systems they studied are *neither* interactively complex nor tightly coupled' (Leveson et al., 2009: 231). The different approaches seem to live peacefully side-by-side, and rather than undermining each other they contribute to expanding both the conceptual space and the scope of safety research. Nevertheless, the divergence of the models may reflect a desire to generalise more than the empirical reality itself justifies.

In the following, a different approach to understanding sociotechnical work and its outcome is outlined. A characteristic of this approach is its particularity and *descriptive* focus and the lack of ambition to develop a generic model/representation of safety and accident genealogy.[4]

5.4 Ethnographic Studies of Centres of Coordination

As shown in Sections 5.1 and 5.2, studies of accidents and safety can be guided by theoretical models of how accidents occur and how they are prevented. An alternative approach to such studies is an ethnographic/ethnomethodological approach, where the emphasis is placed on the observation and description of work as actually performed without comparing it to archetypical models. The data collection and analysis of these studies are often guided by grounded theory (Glaser, 1994; Glaser and Strauss, 1967) where the researcher enters the field without a predefined frame of interpretation. Observations, interpretations and analyses are undertaken in a continuous and open process, and theory is developed rather than applied.

---

4   Not because it is not considered useful, but because it is believed to be impossible and thus potentially misleading and dangerous.

Much research on *centres of coordination*[5] has been undertaken over the last 20 years, and many of the studies have followed this ethnographic tradition. Among the contributors are Suchman (1996b) on airport ground operation rooms, Heath and Luff (1992) on line control centres in the London Underground and Tjora (2000) on acute medical communication centres.

Centres of coordination can be characterised by an ongoing orientation to problems by the participants, involving deployment of people and equipment across distances according to emergent requirements of rapid response to time-critical situations (Suchman, 1993). Some central themes in this research are reviewed briefly in the following chapters.

5.4.1 Awareness

Suchman (1993) mentions some seemingly paradoxical premises for effective performance; an initial division of labour must be combined with the ability of each worker to be aware of their colleague's work in order to make cooperative work possible. In addition, it might be necessary to step in and do the work of a colleague when they are busy with other tasks. A combination and configuration of work processes and artefacts contributes to attaining the necessary awareness.

Centres of coordination are usually designed so that the workers are visible to each other, and it is possible for everyone to see and hear the others' work. Thus, the division of labour is far from absolute. While performing a task, one may also be attentive to what is going on in the room. If a task is believed to be of importance to colleagues, one may render visible the activity or parts of it and encourage others to become sensitive to particular events. Heath and Luff (1992) show how this configuring of awareness is made in the line control centres of the London Underground by, for example, talking aloud in the room, pointing at screens or listening to other's conversations or phone calls. Information can be shared by simply pointing to a screen, work that would have been much more demanding without such an artefact.

5.4.2 Reading a Scene

Suchman (1993) shows how work in an airport control room implies different activities in order to maintain an institutionally accountable spatiotemporal order. One of the themes she focuses on is how a scene is read and understood by juxtaposing and interpreting verbal reports, visual images and various forms of text into an assessment of an emerging situation. In one example, she shows how a Flight Tracker[6] in the operations room is responding to a request from an incoming

---

5   These studies are sometimes labelled control room studies. In more geographically distributed organisations, centre of coordination is a more adequate label than control room.

6   A Flight Tracker tracks arriving and departing airplanes, communicates with pilots on the ground and clears their arrival at designated gates.

pilot. The Flight Tracker seeks information by browsing video monitors, the radio log and the Flight Information Display at the same time as talking with the pilot, in order to get an understanding of the pilot's request. The Flight Tracker is not provided with a single master perspective, but with a range of partial information resources that must be combined into a coherent view.

5.4.3 Structures of Participation

In another example, Suchman (1993) describes a situation where an incoming pilot finds a plane occupying the position reserved for him. The Flight Tracker receives a call about this, and carries out the same actions of browsing through the equipment as in the former example. While they can confirm the status, they are not immediately able to provide the pilot with an interpretation and a solution. However, the Ramp Planner, who is also located in the operations room, overhears the conversation. Without disturbing the Flight Tracker, they are retrieving the same information from the instruments. On the basis of the division of labour, their background knowledge of the situation differs from that of the Flight Tracker, and they are therefore able to read the scene differently. With a short sentence, the Ramp Planner informs the Flight Tracker of the situation, on the basis of which they in turn can inform the pilot. In this example, a combination of the reading of a scene, lending eyes and ears to a colleague's work and interjecting without being asked are techniques that shape the structure of participation that enters into the coordination of work.

5.4.4 Technologies as Material Practice

The role of artefacts and ICT (information and communication technology) systems for providing several layers of information is another theme elaborated on by Suchman (1993). Not only can a combination of ICT contribute to the reading of a scene, but the fact that the same information is accessible to several workers makes a shared awareness possible without a great deal of communication between them. It is also argued that the technologies cannot be separated from the activities of their use. The functionality of technological artefacts is thus not located in particular devices, but in the 'structured courses of action involving the assembly of heterogeneous devices into a working information system' (Suchman, 1993: 44).

5.4.5 Shared Understanding[7]

While situated action, awareness and the reading of scenes certainly address important issues of cooperative work, they do not address the challenges that arise

---

7   There are many other terms describing aspects of the same phenomenon, e.g. common operational picture and shared situation awareness.

when people are cooperating at a distance, as is the case in Integrated Operations. When workers are not co-located, they cannot clarify interpretations of information face-to-face, as in traditional control-room settings (Bannon and Bødker, 1997).

Creating shared views, which is a major goal in the control-room settings described above, is only part of the issue addressed in more scattered systems. In IO, where people and instruments may be distributed over large areas, creating shared understanding is also a challenge. Even with access to the same representations, the development of a shared understanding involves different, and maybe also more, articulation work (Strauss, 1985, explained below) than when workers are located in the same room.

Although shared understanding as an enabler of successful collaborative work is a recurrent theme in IO (e.g. Andresen et al., 2006; Grøtan et al., 2009; Hepsø, 2006; Rolland et al., 2006; Skarholt et al., 2008; Tinmannsvik, 2008) and the breakdown of shared understanding is often used to explain accidents, the role of shared understanding in cooperative work should not be taken for granted, as discussed in Chapter 9.

5.4.6 Articulation Work

In the field of computer-supported cooperative work (CSCW), articulation work refers to the work of coordinating, aligning or integrating interdependent activities (Schmidt, 1996; Strauss, 1985), i.e. a set of activities required to bring together discontinuous elements into working configurations (Suchman, 1996b). Articulation work can be understood by contrasting it to primary work. While primary work is 'the activities that directly address the specific agendas and goals of the work situation' (Hepsø, 2006), articulation work is less standardised and formalised, it is 'work that gets things back "on track" in the face of the unexpected, and modifies action to accommodate unanticipated contingencies' (Star and Strauss, 1999: 10). Articulation work often has a low visibility in rationalised models of work, and thus represents a category of activities that are at risk of being sacrificed – intentionally or unintentionally – during change processes. Articulation work receives specific attention in Chapter 8.

5.4.7 Situated Action

A common feature of the above themes is the connection to situated action. Suchman (1987) introduced the term situated action to the field of sociotechnical work studies. As an alternative to the interpretation of work as a set of well-defined goals that people plan ahead of time, in Suchman's perspective 'every course of action depends in essential ways upon its material and social circumstances' (Suchman, 2007: 70). This is a keynote in the concept of situated action, and it is an important perspective in many studies of centres of coordination. The role played by plans is not primarily as scripts for how action should be performed, but rather as resources for action and post hoc representations and rationalisation of work.

5.5 Actor-Network Theory

The above review of genealogic, model-based[8] and ethnographically based understandings of accidents and safety describes perspectives that could also be distinguished as generic and holistic frameworks on one hand, and singular and unique descriptions on the other.[9]

In this investigation of offshore operations it became difficult to produce descriptions and explanations without the support of a more 'flat' methodology that could transcend the borders between the micro and macro levels, between local interaction and theoretical concepts, and between the social realm of humans and the material realm of artefacts. The controversy between the NAT approach and the HRO approach, for example, cannot be overcome without an epistemology that treats humans and artefacts symmetrically.[10] Systemic models and individual actors cannot enter into the same descriptions unless the language of explanation allows the researcher to move freely between local interactions at the micro level and theoretical concepts of the macro level.

Rather than a theory, Actor-Network Theory (ANT) (Latour, 2005) could be conceived as an epistemology and a method. In ANT, what is usually divided into epistemology and ontology collapses into a constructivist understanding of how the world is constructed and how knowledge about it can be achieved. The usefulness of such a constructivist perspective for understanding sociotechnical systems has been demonstrated in a range of studies (e.g. Callon, 1986; Latour, 1988; Law, 1986).[11] In studies of accidents and industrial safety, however, it has not yet been much applied (but see e.g. Law, 2003a, b)). Some central themes of ANT that are important for building the arguments of this book are briefly reviewed in the following chapters. For those who do not wish to go deep into this theoretical perspective, the rest of this chapter will nevertheless be a useful resource of vocabulary for the rest of this book.

5.5.1 Mediation

Mediation is a useful term when *actors* are to be distinguished from *intermediaries*. Whereas intermediaries have a fully predictable effect on a change process in the sense that the output can be predicted from the input and the intermediaries, actors – or mediators – are recognised by the unpredictable difference they make. They

---

8   Describing HRO as model-based is open to criticism. However, it may be defended by the following argument: the original HRO research was perhaps ethnographic, but it has to some extent been closed and the results put to use as a prescriptive model.

9   This is admittedly a coarse classification that may not fit all work from those traditions.

10   The topic of symmetry is elaborated on later.

11   Constructivism is a term with many meanings and connotations. Read Section 5.5.6 for clarifications of the way the term is used in this book.

are not placeholders through which causality can be transported; they are change agents that mediate the other actors with which they interact (Latour, 2005).

This distinction between mediators and intermediaries is central in ANT. It contributes to nuance the discourse of causality affected by abstract entities such as culture or power; if culture or power is predictably transported through any agent, that agent should not be considered as an actor. If, on the other hand, culture or power is affected in any way by an agent, that agent is a mediator, an actor.

Careful application of the terms mediators and intermediaries makes a big difference when investigating how the introduction of new technologies and work processes into an existing sociotechnical system may affect the system. One implication is that the investigation is more oriented towards *relations between actors* than towards *properties of components*, since it is at the meeting between the actors that changes occur, not at the designer's drawing table.[12]

5.5.2 Heterogeneity and Symmetry

Empirical descriptions guided by ANT may be quite complex. Mapping an actor-network involves mapping elements of different constitutions. In his study of the introduction of an electric car in France, Callon (1987) documents what the associations of such different elements may look like; a network of electrons, batteries, social movements, industrial firms, ministries and consumers. Acknowledging this heterogeneity and acknowledging that 'objects too have agency' (Latour, 2005: 63) is vital to make trustworthy accounts of sociotechnical change processes. The imperative symmetry in such accounts implies that the accounts result from one single repertoire used for describing both social and material elements. This means that as ANT researchers investigating sociotechnical systems, '[we] should not ... change registers when we move from the technical to the social aspects of the problem studied' (Callon, 1986: 200).

Thus, analyses of sociotechnical systems such as integrated drilling operations guided by ANT differ from traditional analyses that seek to account for the heterogeneity by applying holistic approaches of MTO.[13] While most analyses performed by engineers or sociologists rely on separate analyses of predefined components of MTO, and thereafter look for how they affect each other, analyses guided by ANT will not let such a priori categories determine the direction of the research. The practical meaning of heterogeneity and symmetry is demonstrated in Chapter 7.

---

12   But the designer is indeed also a potentially powerful actor, pertaining to how clearly they are able to inscribe their programme into a technology or work process.

13   Norwegian: '*Mennesker, teknologi, organisasjon*' (people, technology, organisation).

5.5.3 Complexity

Although complexity is a recurrent theme in safety research, the term has different meanings in different research communities and is therefore a problematic entity. A distinction may be drawn between ontological and epistemological complexity (Hollnagel, 2008a). Ontological complexity refers to the complexity of the 'real' system, whereas epistemological complexity refers to the complexity of the system *description*, i.e. how the system is perceived. The complexity in NAT is of an ontological nature, whereas resilience engineering seems to refer to an epistemological complexity.[14] However, the criteria by which these complexities are assessed are not clearly defined.

Latour's (2005) view on complexity differs from both these definitions, because complexity is seen as the inverse of the resources available for maintaining order in the system.[15] Thus, lower complexity is acquired with additional material or symbolic resources that may be used to stabilise, and if possible black box, a sociotechnical system. This could seem counter-intuitive, as the general comprehension of complexity is that the more elements added to a system, the more complex the system becomes. According to ANT, more elements may make the system more *complicated*, but not necessarily more complex. However, the stability may be fragile; a properly functioning computer is an example of a complicated, black-boxed artefact that may become terribly complex in the case of malfunctioning (Latour, 2005).

When evaluating the consequences of introducing new technologies and work processes in drilling operations, such a definition of complexity will inevitably make the evaluation more ambiguous and less stable than the above referred ontological and epistemological definitions, because the complexity cannot be evaluated without accounting for the context into which it is introduced.

5.5.4 Black Boxing

When an innovation is in the making, it is often full of controversies and its functionality and impact is still not unambiguously defined. Many central writings in science and technology studies (STS)[16] (e.g. Bijker, 1995; Latour, 1988, 1996a; Shapin and Schaffer, 1985) have resulted from studies of science *in the making*, i.e. before it was black boxed.[17] The notion of the black box is borrowed from

---

14   Hollnagel (2008a) prefers the term intractability to complexity.

15   See Strum and Latour (1987) for a fascinating and illustrative elaboration of the complexity of social order among baboons.

16   STS is an 'interdisciplinary field that is creating an integrative understanding of the origins, dynamics, and consequences of science and technology' (Hackett et al., 2008: 1). ANT may be seen as one of several 'schools' developed from STS.

17   Indeed, many studies of innovations in the making are retrospective studies, but through the use of numerous sources retrospective studies are also capable of observing and

cybernetics, where it denotes a device about which the users only need to know the input and the output. In ANT, it is a label that is put on phenomena (technologies, work processes, etc.) after their internal controversies have been silenced and their functionality and comprehension has become unambiguous (Latour, 1987).

This book investigates a change process that is still under development and thus has not yet been black boxed,[18] as well as concepts and work forms that have already been black boxed and which therefore must be reopened in order for their constitution to be explored.[19] At a higher level the concept of Integrated Operations is itself a concept in the making. Although it is not difficult to locate controversies that are connected to IO, many actors are eager to *close* the concept and to associate it unambiguously with faster, better and safer operations. Section 2.1 and Chapter 8 describe an inspection into a work process that is black boxed to most outsiders.

5.5.5 Inscriptions and Immutable Mobiles

Inscription is a general term that refers to 'all the types of transformations through which an entity becomes materialized into a sign, an archive, a document, a piece of paper, a trace' (Latour, 1999b: 306) – an immutable mobile. The power of immutable mobiles can be illustrated by the impact of maps on modern geography (Latour, 1986). Through a series of transformations, the resulting inscriptions become mobile.[20] These inscriptions also become more immutable than the original substance or representation, because their properties (spatial dimensions, colours, weights, etc.) may be kept stable in the form of paper inscriptions.

Latour and Woolgar (1986) illustrated the power of inscriptions when they documented the work processes at The Salk Institute for Biological Studies. In contrast to how one may usually think of a laboratory, it was depicted as a factory of paperwork; a place where various items of apparatus 'transform pieces of matter into written documents ... a curve or a diagram' (Latour and Woolgar, 1986: 51) that in turn will constitute far more powerful arguments in the technical or scientific debates than arguments based on reference to the original substances, especially when they are combined with other inscriptions. Combinability is thus another important characteristic of inscriptions.

---

describing a process in the making.

18   For example, tools for monitoring, diagnosing, automating and visualising the drilling process.

19   For example, shared understanding.

20   Consider the difference in mobility between a land area and the map that is the end product of a comprehensive transformation process.

5.5.6 (Social) Construction

While the first edition of *Laboratory Life* (Latour and Woolgar, 1979) had the subtitle *The Social Construction of Scientific Facts*, the edition published in 1986 was simply titled *Laboratory Life. The Construction of Scientific Facts* (Latour and Woolgar, 1986). The word 'social' was omitted due to the many misreadings of *social construction*. However, the omission did not reflect a view that the construction of facts is not *social*. But then again:

> [H]ow useful is it when we accept that all interactions are social? What does the term 'social' convey when it equally refers to a pen's inscription on a graph paper, to the construction of a text and to the gradual elaboration of an amino-acid chain? (Latour and Woolgar, 1986: 281)

A precise definition of 'the social' makes a big difference to the understanding of how 'construction' is conceived in ANT. 'The word "social" … does not designate a "kind of stuff" that distinguishes it from other types of materials, but the process through which anything, including matters of fact, has been built' (Latour, 2003b: 28). What makes the construction 'social' is 'the many heterogeneous ingredients [workers, architects, masons, cranes, concrete etc.], the long process, the many trades, the subtle coordination necessary to achieve such a result' (Latour, 2003b: 29). That a scientific fact is socially constructed does not mean that it can be easily deconstructed because it is made of 'social stuff'; the more heterogeneous the actors involved in the constructed work-net[21] (Latour, 2005: 143), the more solid and resistant is the construction. This understanding of social and construction is necessary in order to distinguish between non-modern and post-modern understandings of social construction (Latour, 1993).

5.5.7 Controversies

Studies of scientific and technological work have shown how the progress of work often is accompanied by black boxing of the controversies involved in the work (Latour, 1987). One effect of this black boxing is that the result of the work appears clear and unambiguous, and is thus easily comprehensible and applicable to those who depend on it. To those who study work, however, black boxing obscures the biography of the end product and thus the controversies that enter it. No matter how reconciled these controversies are, they may still be influential constituents of the resulting product that may again become relevant if the conditions are right (or wrong!). Whether one speaks of the DNA molecule having the shape of a double helix (Latour, 1987), of the determination and synthesising of the thyrotropin-releasing hormone (TRF) (Latour and Woolgar, 1986) or the depth of the Garn

---

21 Latour sometimes writes work-net instead of network to underscore the connotations of movement and changes rather than static configurations.

formation in the North Sea (see Section 2.2 and Chapter 9), there is more to learn from the process of constructing the fact than from the mere end product. This is why Latour recommends studying *science in the making* rather than *ready-made science*: 'We study science in action and not ready made science and technology; to do so, we either arrive before the facts and machines are blackboxed or we follow the controversies that reopen them' (Latour, 1987: 258). The usefulness of adopting such an approach in this study is especially evident in Chapters 6 and 10.

5.5.8 Centres of Calculation

The inscriptions and immutable mobiles referred to above may be combined with other inscriptions collected from other sources. Such combinations are especially effective where models and equations exist that suggest how the different inscriptions relate to each other and how they should be superimposed to produce yet more refined information, and tools that can perform those combinations quickly and accurately. Latour (1986, 1987) calls such facilities *centres of calculation*. In centres of calculation, heterogeneous elements that in their more original appearance are very hard to combine arrive as representations (numbers, graphs, etc.) that are possible to combine using such equations and models. When Edison worked on the development of an electricity infrastructure he had to calculate how much copper he needed. For such calculations, he had to combine information from very different domains.

> By manipulating the equations, [Edison] retrieves sentences like: the more you increase the cross-section to reduce loss in distribution, the more copper you will need. Is this physics, economics or technology? It does not matter, it is one single web that translates the question 'how do you bring down the price of copper' into 'how can you fiddle with classic equations of physics'. (Latour, 1987: 240)

Access to such centres and tools is a prerequisite for utilising a large amount of inscriptions stemming from different sources. It also renders comprehensible the close relationship between cognition and coordination indicated by Latour (1986, 1999a) and Hutchins (1995a). Equipped with tools that continuously collect drilling data and combine and evaluate them in accordance with real-time calibrated theoretical models of the drilling process, the distributed cognitive capability of both onshore rig teams and offshore rig crews may improve dramatically if the teams and crews are able to wisely manage and arrange the tools and inscriptions.

5.5.9 Actors and Relations

A central perspective in ANT is that action, by definition, is never performed by one actor alone. Borrowing a metaphor from Goffman, Latour emphasises that once the play starts, nothing is certain with respect to the origin of action:

Does the audience's reaction count? What about the lighting? What is the backstage crew doing? Is the playwright's message faithfully transported or hopelessly bungled? Is the character carried over? And if so, by what? What are the partners doing? Where is the prompter? If we accept to unfold the metaphor, the very word actor directs our attention to a complete dislocation of the action, warning us that it is not a coherent, controlled, well-rounded, and clean-edged affair. By definition, action is *dislocated.* (Latour, 2005: 46)

Action should thus be searched for in *associations*, in the network of heterogeneous actors, in the relations rather than in the properties of the different actors. A variant of this advice may be found in Chapter 7, where it is argued that the outcome of events cannot be deduced from the qualities of singular components, because it results from unpredictable[22] interaction between actors.

5.5.10 Following the Actors in a Flat World

To follow the actors is a core principle of ANT (see e.g. Latour, 1987, 2005). It may seem a rather mundane principle, but taken seriously it is a very demanding principle that turns sociology into *slowciology* (Latour, 2005). *Laboratory Life* (Latour and Woolgar, 1986) offers an example of how it can be done in practice. Rather than deciding in advance what to look for when the work of the scientists in the laboratory is to be studied, the authors follow the scientists and write down accounts of their activities; the connection with other actors, the mediation of chemical substances into graphs and reports, the translation of others' interests and the publication of papers.

Such a modus operandi presupposes a *flat earth* in the sense that no a priori discrimination is made between agencies and structures, or local interaction and global context. To explain what this means, Latour elaborates on the scientists of a flat earth:

You talk to them about a company managerial structure – terms that everyone understands easily (I mean the 3-D people) – and those 2-D analysts will ask you about the office space in which one manager sits, then the book-keeping procedure she uses, then the chart on which she has labelled the names of her collaborators, then the chain of command which allows her to call for meetings, then the pie-chart which she has received from marketing and that describes consumer response to her latest campaign. Do these statistics measure the demand for the product? Does the organization chart reveal the corporate structure? Not a bit, and that is what is so terrible with them. Those instruments do not measure anything, they make the things they measure! (Latour, 1996b: xii)

---

22   This is not to say that every system is unpredictable. Systems that are completely closed and automated may show nearly full predictability. However, where there are actors – in the sense of mediators – the outcome of interaction can never be fully determined.

This quote is provided in such length not only because it sets the focus on how to follow heterogeneous actors in a flat world, but it also sums up Section 5.5 by illustrating how ANT scientists may practically approach their sociotechnical study objects.

5.5.11 Translations

Action is difficult to isolate to single actors, single events or social or technical structures when human and non-human actors are treated as mediators and not simply vehicles to carry an effect forward (Latour, 2005: 58–59). The way action is understood in the ANT literature is as a chain of events involving *translations*. Translation is a term with both linguistic and material connotations, and it refers to the process through which actors mediate and displace representations and interests (Latour, 1999b). A specialised meaning that defines the term in a way that is particularly relevant to this study[23] is that a translation is 'a relation that does not transport causality but induces two mediators into coexisting' (Latour, 2005: 108).

To understand how the concept of translation may be used to describe chains of events, the mundane example of how the shape of a door key may change both a hotel manager's desire and his guests is provided (Latour, 1991): The *programme* of the hotel manager is to get the hotel guests to leave the key at the reception when leaving the hotel. This programme is first translated into an oral message that is communicated to each guest. When this does not change the behaviour of all the guests, the programme undergoes a second translation into a written message that is distributed to all guests. However, there are still some hotel guests that, for different reasons, still practice the *antiprogramme*;[24] keeping the key when they leave the hotel. A third translation adds a metal weight to the key and thus makes it heavy and uncomfortable to carry in the pocket. The strength of the programme is now sufficient to convince the majority of the hotel guests to follow the manager's programme. The translations have altered both the desire of the hotel manager and the behaviour of the guests. In the beginning, the manager's desire was naked, in the end it was clothed, or loaded. Or as Latour says: 'In the beginning it was unreal; in the end, it had gained some reality' (Latour, 1991: 108). In the beginning, the majority of the guests took their key with them when they left the hotel; in the end most of them deposited it at the counter (see Figure 5.9).

The concept of translations will later be applied to describe change processes in the concrete sociotechnical system of a drilling operation/organisation.[25]

---

23   I am here especially referring to Chapter 7 and the discussion in Section 11.1.

24   The theoretical argumentation differs, but the practical content bears a clear resemblance to Turner's anti-tasks, as referred to in Section 5.2 and in Turner (1978).

25   It is especially relevant in the discussion in connection with Chapter 7. See also Hepsø (2009) for a more comprehensive elaboration on the issue of translations in the petroleum industry.

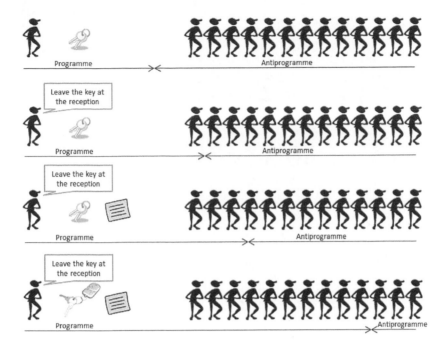

Figure 5.9   Translations may strengthen a programme of action

Source: figure inspired by Latour (1991)

Chapter 6
Methodological Themes

6.1 *In Medias Res*

Setting a starting point and an end point for the drilling operations, to separate them from the foregoing and consecutive operations, is a pragmatic act that must be done with careful reference to the objective of the study. One *could* say that the operations start when the drill string is first run into the geological formations – but then the preceding process of planning would be neglected. One could also say that the drilling operation is finished when the well is completed and the production of oil or gas starts – but in that case one disregards the actual value of production data to later drilling operations. In this study, this dilemma is solved by *not* undertaking any a priori demarcation with respect to location, organisation chart, or time window. Instead, what is chosen as the point of departure is a drilling *situation* combined with a drilling *perspective*. Adopting such an approach implies that the focus of the study is defined by the changes that the well undergoes, and the heterogeneous actors that affect and are affected by them.

The onshore rig teams[1] define the perspective from which the operations are studied. The rig teams are responsible for the construction of the wells, all the way from planning to completion. What is sought is a position from which most aspects of drilling operations can be observed and comprehended, and admission to a rig team thus locates the researcher right at the centre of the scene.

A well that was under construction at the time I was granted admittance to the field came to serve as the situation from which the unfolding of events and the tracing of networks could begin. This part of the fieldwork extended through most of the of the well construction period, although it started after the planning phase and ended before the completion of the well. It contributed to a concrete case that is treated in Section 2.1 and Chapter 8, and, more importantly, it functioned as an entry into the practical work of drilling. Typical actors and artefacts contributing to the operations were identified, and a valuable context for the future interviews and the IO tools was provided through the observation of the rig team's work.

---

1 Throughout the book, 'rig team' refers to the *onshore* group responsible for planning and follow-up of the drilling operations, while 'rig crew' refers to the *offshore* group carrying out the operations.

6.2 The Devil is in the (Normal) Details

In a high-risk activity such as offshore drilling operations, one might expect a study to reveal conditions and incidents of a risky and perhaps spectacular character. This would certainly be an adequate expectation if the study followed a tradition where safety is seen as the absence of risks and errors. However, this is a safety perspective that has been challenged in recent years by such approaches as HRO (La Porte, 1996; La Porte and Consolini, 1991; Weick, 1987; Weick and Roberts, 1993; Weick and Sutcliffe, 2007) and resilience engineering (Hollnagel et al., 2008, 2006; Nemeth et al., 2009). Both these approaches aim at describing the conditions that lead to safe performance rather than the conditions that lead to accidents. Hollnagel reasons as follows when he justifies such a focus: 'Even if the probability of failure is as high as 10^{-4}, there are still 9,999 successes for every failure, hence a much better basis for learning' (Hollnagel, 2009: 83), and he adds that 'Studying why things go right will not only make it easier to understand why things go wrong, hence lead to improvements in safety, but it will also contribute to improvements in efficiency, productivity and quality' (Hollnagel, 2009: 101).

In this study, safety is understood as an entity that is inseparable from the work itself. Thus safety cannot be obtained by draping some kind of primary work in a new layer of secondary HSE[2] work. Since both safety and accidents are mainly the results of variations of normal work performance – variations that are impossible to avoid if the work is to be done in a practical manner (cf. ETTO, Hollnagel, 2009) – this normal work should be studied in order to understand the qualities of work that actually produce safety (or accidents).

It follows from this argumentation that interesting observations do not have to involve deviances from procedures, dangerous situations or accidents to produce significant insights about the preconditions for safe and successful work. On the contrary, it is possible to discover details that are *necessary to make work work* in mundane, invisible everyday activities (Schmidt, 2010). Similarly, it may be found that conditions that are *thought* to be necessary to make work work may not be so essential after all. States of both these affairs are elaborated on in Section 2.1 and Chapter 8, and Section 2.2 and Chapter 9 respectively.

Another advantage of studying normal work is that it is easier to avoid predisposed descriptions and explanations. Many models or frameworks of accident genealogy exist, and when these are applied to explain why accidents occur there is always a risk of tautological reasoning. One such example could be the case of *shared understanding*: if it is believed that breakdown of shared understanding qualifies as a reason why an accident may occur, accident investigations may reach conclusions on the cause of accidents if they identify a breakdown of shared understanding prior to the accident (Dekker and Hollnagel, 2004). In a certain case this may be right or it may be wrong – the important thing is that it skips the process of unpacking the *concept* of shared understanding and the investigation of

---

2   Health, safety and environment.

what is shared and what is not shared between which actors, and *in which way* this *affect* the progress of work.

When studying normal work, fewer models and frameworks to support the descriptions exist.[3] This makes the task complex and time-consuming, and one might not know what to look for, or what resources it will require to find it. On the other hand, the risk of tautologies and predisposed explanations misguiding the descriptions is much smaller. This does not only relate to the researcher's observations and explanations. A 'negative view on safety'[4] may also influence the informants' descriptions.

Chapters 8 and 9 illustrate that important lessons can be learned from studying 'normal work'. The important thing is that the mundane details that may be observed in such studies are very hard to find and to make sense of in studies that focus on risks and errors.

6.3 Interviews, Observations and Literature Studies

Interviews were undertaken during the period from December 2008 and through December 2009, in bouts of varying intensity. In some periods several interviews were undertaken the same and/or following days in accordance with a planned sequence, while at other times they were fewer and more ad hoc, related to specific questions arising during observation studies or during writing. Roughly half of the interviews were tape recorded and later transcribed.[5] Swift notes were taken during the rest of the interviews, followed by more thorough written accounts as soon as practically possible after the interview situation.

Many of the informants for the interviews were identified through a snowball recruitment process. During the interviews the informants were asked to think of other candidates who might throw light on the topic. Their suggestions were most often supplemented with contact information and a personal greeting to encourage the potential informants to participate. Other informants were recruited more directly during the observation studies. Some of the involved participants in these studies were also interviewed, either in between the study sessions or at a later point in time. In addition, I had access to a series of interviews

---

3　There are, of course, the work procedures produced by the institution, but using these for identifying discrepancies brings us back to the discourse of compliance, errors and accident genealogy models described previously.

4　Focus on why things go wrong rather than why things go right.

5　It would have been desirable to tape record and transcribe all interviews. Whether or not an interview was tape recorded was a matter of judgement in each case. In some cases, the informant did not want the interview to be recorded. Other interviews happened spontaneously when the recorder was not available, and yet other interviews were not recorded because it was believed that this would have a negative impact on the interview situation in one way or another.

performed by colleagues in connection with another project on the experiences and expectations related to a specific IO tool (Letnes et al., 2008). Although these interviews were not tailor-made for this study, and thus less directly used in the book, they constituted valuable background information. Thirty-four informants were interviewed. An additional 18 transcribed interviews originated from the project mentioned previously.[6] For an overview of the informants and their professions, see Table 6.1.

The interviews were qualitative and open and were undertaken in an informal manner. An interview guide was initially developed, but this came to be used merely as a tool for reflection prior to the interviews, taking the form of notes and ideas as a part of the preparations.

The goal was not to confirm or disprove any established hypotheses, but to explore the field and topic as perceived and enacted by the profession itself. This exploratory approach was intended to open up the field rather than to narrow it down – to trace the work-net (Latour, 2005: 143) and the way it might influence and be influenced by new technologies and work processes.

Table 6.1   Interviews – informants and professions/positions[7]

Informants' professions	Number
Drilling engineers	9
Drilling plan leaders	4
Drilling superintendents	5
Drilling data analysts	3
Geologists	5
Reservoir engineers	4
IO advisers/managers	3
IO technology developers	10
Researchers drilling and well	4
Service company	1
IT-support	1
HSE engineer	1
Total	**50**

---

6   Two informants were interviewed in connection with both projects, thus 50 informants gave 52 interviews.

7   All the informants have their primary workspace in onshore offices. However, the drilling engineers, the drilling plan leaders and the geologists travel offshore on a more or less regular basis. In their earlier careers, most of the informants have worked on offshore rigs.

The choice of cases for the observation studies was motivated by the research questions and the organisational arrangement of the work processes of offshore drilling operations. To explore the influence of new tools and work processes on the safety of the operations, the best strategy was to study the actors and the places through which most information circulates, where advice is sought and decisions are made. Thus a subsurface expert centre and an onshore operations centre were contacted.[8]

I spent about three months at the onshore operations centre. Having my own dedicated workspace[9] in the open-plan office, I could come and go as I wanted. For some periods I stayed together with the rig team the whole day and for several days in a row, while during other periods when activity was lower or I needed time to review my field notes, I paid shorter visits only to observe events I considered especially important (e.g. meetings with onshore partners or with the offshore crew). A concrete output from this part of the field work was the case presented in Section 2.2.

The study of the subsurface support centre was shorter and more intense. I spent two full working days with the team, during which time I was allowed to follow a process of problem solving from beginning to end. This specific case forms the basis for Section 2.1 and Chapter 8.

These two locations for observation studies were, in this study, regarded as two of three core locations – the third being the offshore rig – for the accomplishment of drilling operations. The combination of studying these, having the professional work experience from offshore rigs and undertaking interviews with a broader spectrum of employees – see Table 6.1 – was considered to be adequate for the task. In addition, literature and documentation on integrated work processes and technologies, and conferences and forums[10] focusing on IO provided valuable sources of information.

---

8   A third alternative would be the offshore crew. Surely important lessons could have been learned from such a case study. The reason why it was ruled out was twofold: first, time constraints of the study and many more administrative obstacles for access would have made it much more challenging. Second, since my former work experience is from the offshore location, I already have knowledge of this work and have more to gain from visiting the onshore locations, of which I have less knowledge.

9   This workspace belonged to the team's drilling engineers who were offshore at the time. Thus, I only borrowed the workspace until this engineer returned onshore and I had to switch to the workspace of the next drilling engineer who travelled offshore.

10   For example the Conference on Integrated Operations in the Petroleum Industry (www.ioconf.no) and the Center for Integrated Operations in the Petroleum Industry (www.ntnu.edu/iocenter).

6.4 An Ethnographic Inspired Approach

The approach of this study is inspired by ethnographic workplace studies, which in turn are indebted to the ethnographic tradition of anthropology. Many sources of inspiration (e.g. Engeström and Middleton, 1996; Knorr-Cetina, 1981; Latour and Woolgar, 1986) have been of great help as they have cleared the way for a practice for social studies of science beyond the Cartesian orthodoxies (Engeström and Middleton, 1996).[11] Latour and Woolgar (1986: 279) write that 'the anthropologist *does not know* the nature of the society under study, nor where to draw the boundaries between the realms of technical, social, scientific, natural and so on', and this also a guiding star in this book. They continue their reasoning – thus further describing the perspective and the methodology of the present study – by describing their working principle:

> This additional freedom in defining the laboratory counts for much more than the artificial distance which one takes with the observed. This kind of anthropological approach can be used on any occasion when the composition of the society under study is uncertain. It is not necessary to travel to foreign countries to obtain this effect, even though this is the only way that many anthropologists have been able to achieve 'distance'. Indeed, this approach may very well be compatible with a close collaboration with the scientists and engineers under study. We retain from 'ethnography' this working principle of *uncertainty* rather than the notion of exoticism. (Latour and Woolgar, 1986: 279)

The fact that field observations will always be 'contaminated' by the presence of the observer, and that the 'natural' state of the field can never be 'objectively' described, was not regarded as a problem in this study. On the contrary, it was a specific aim to interact with the informants in such a way that their actions and their answers reflected their responsibility for their profession and their work, not to the researcher and the research project. The biggest risk was not considered to be a lack of objective descriptions and explanations due to influence from the researcher, but that the informants would be too compliant with the researcher and with existing discourses on the prospective influence of the industrial change process of Integrated Operations. The measure of objectivity was not disinterest and absence of the researcher, but rather the informants' *recalcitrance* (Latour, 2000) and their opportunity to *object*.

---

11   Social studies of science beyond Cartesian orthodoxies means to 'investigate work activity in ways that do not reduce it to a "psychology" of individual cognition or a "sociology" of communication (whether "micro" or "macro") and societal structures' (Engeström and Middleton, 1996: 1).

6.5 Making Sense of the Data

The data analysis is dominated by three main approaches. In Chapter 7, a case and a hypothetical scenario is used to facilitate a discussion of the nature of sociotechnical systems. As the basic ontological perspective that constitutes the fundament for the whole study is already set,[12] the analytical task is mainly to search the empirical material for arguments for and effects of this ontology. The analysis thus represents the first test of the validity of this ontological perspective in the field of inquiry, and its suitability to generate new knowledge about the field. The analytical method was to identify instances in which the actors involved are mediated through the course of events and to elaborate on the dynamics of the relationship between the actors. This involved a re-reading of the original case presentation and analysis where less attention was paid to the original framing of the empirical material into different dimensions, or organisational attributes, and more to the point of mediation, accounting for the events in which the actors' competencies are acquired and altered.

The analytical procedure in Chapter 8 and Chapter 9 is more descriptive and more explorative than that in Chapter 7. The analyses are not directed towards supporting any pre-stated argument or hypothesis, but rather towards describing how cooperative work processes are both made possible and constrained by the heterogeneous character of the involved actors. The heterogeneity refers to differences not only in constitution and competencies, but also in objectives and success criteria. Importantly, the analyses of these two cases did not take place *after* the fieldwork was carried out, but were carried out continuously from the start of the field work to the completion of the writing. Through a hermeneutical process of interpreting the empirical observations and letting the interpretations guide the direction of the fieldwork, the analysis has been intertwined with the empirical fieldwork itself.

Chapter 10 makes use of a third analytical strategy. The empirical data consists of interviews, most of which were transcribed. Here, the analytical process to a much larger extent succeeded the fieldwork; it was separated from the field both geographically and in time. The analysis involved a process of coding that was inspired by the emergent coding of grounded theory (Glaser, 1994; Glaser and Strauss, 1967); the first reviews of the empirical material resulted in topics that in turn were compared and grouped so that a smaller set of themes appeared. These themes then guided the subsequent reviews, and through this alternation between the empirical material and the interpretation of it the analyses resulted in

---

12   In short, this ontological perspective implies that in a study of a real-world phenomenon, a model of the world/phenomenon, the actors populating it and any driving forces that produce the outcome do not enter into the starting conditions and the methodological resources of the study. On the contrary, these conditions are considered a *result* of the phenomena that are described. See also Section 5.5.

the understanding of how the research question could be answered with reference to a set of controversies identified in the field.

The descriptions and interpretations throughout the book are of a qualitative nature. This is a consequence of the objectives and the study's explorative form. This does not mean that quantitative methods could not prove useful in later studies where the objective is operationalised in such a way that quantitative support is found adequate and desirable.

A few informants became important to the study beyond their mere contributions to the fieldwork because they were willing to read, comment on and discuss texts as they were produced. This gradually became part of the strategy to ensure not only the scientific *quality* of the work, but also the *relevance* to the science and practice of their field, as seen from *their* professional perspective.

6.6 Truth or Dare

Every method has its strengths and weaknesses. As the epistemological perspective of this study assumes a constructivist and relativist view rather than a correspondence theory of truth, the method is directed towards *matters of concern* rather than *matters of fact*. Relativity is sometimes confused with *anything goes*, but there is indeed a difference between the *relativity of truth* and the *truth of relation* (Deleuze, 1993). The study should thus not be evaluated for its correspondence to the truth, but through trials of strength. This also hints at a possible weak point of the method: the trials of strength carried out by the author could have been more extensive. In other words, more and deeper empirical investigations could have been undertaken. Much of the argumentation is based on cases, and generalisation of cases is always problematic.[13] The intention has been, however, to study typical but specific work situations and to describe them in a way that presumptively has generic relevance. It is a daring approach because it may be subject to criticism from the established traditions within both natural and social sciences. The strength of the method, however, is that it seeks to take into account all the entities, but also only those who mediate or are mediated through the course of events without assuming the landscape of actors or their qualities in advance.

---

13   It is nevertheless the author's meaning that such demurs should not scare the researcher from being empirical. On the contrary, better empirical investigations are necessary for strengthening argumentative power. From the epistemological realists one has to dare to withstand accusations that argumentative power is the expression of social construction and not the truth, and from social constructivists one must dare to withstand accusations of empiricism.

6.7 On Theory, Method and Epistemology

Although the name Actor-Network Theory clearly states that ANT is a theory, it could as well be called a method. Latour actually does so, stating that ANT is 'a method to describe the deployment of associations, like semiotics is a method to describe the generative path of any narration' (Latour, 1996c: 9).[14] The focus on empirical studies to describe the networks – or work-nets (Latour, 2005: 143) – that human and non-human actors produce as they proceed with their work, and the advice to account meticulously for all the translations and mediations that occur, makes ANT a method of description. However, ANT could also be seen as an epistemology. A red thread through all ANT writings is the perspective on what knowledge and truths consist of and how they can be acquired, or rather constructed. This epistemology – in which the traditional distinction between what we know about the world and how this knowledge is constructed (epistemology) and what the world really consists of (ontology) collapses – is nevertheless inextricably linked to the scientific method.

The HRO (High Reliability Organisation) research also oppose to a purely theoretical label. Although sometimes labelled HRT (High Reliability Theory) and widely referenced together with other theories of safety such as NAT (Normal Accidents Theory) and RE (Resilience Engineering), HRO is not really a theory. La Porte and other influential HRO researchers were quite clear that they were conducting case studies on organisations operating extremely hazardous technologies with an extraordinarily good safety record (La Porte and Consolini, 1991), and that they aimed at describing the features that these highly reliable organisations had in common. It was not aimed at making prescriptions for safe organisations, or at constituting a theory of how to become an HRO.

Faced with such approaches that do not easily fit into categories of theory *or* method *or* epistemology, it seemed more fruitful to discuss them in terms that allowed the scientific terminologies to be resources for precise description rather than a rigid structure into which to fit the description inadequately. Through this chapter, some important consequences for the chosen methodological approach have been elaborated on in terms of themes that shed light both on the premises and the results of the study.

In Figure 6.1 the different approaches to understanding safety that have been reviewed are recapitulated and categorised somewhat coarsely in order for the reader to see more clearly the difference the methodological approach makes when in the next chapters we dig deeper into the empirical world and challenges of offshore drilling operations.

---

14   That is not to say that he does not also call it a theory: 'It's a theory, and a strong one I think' (Latour, 2005: 142).

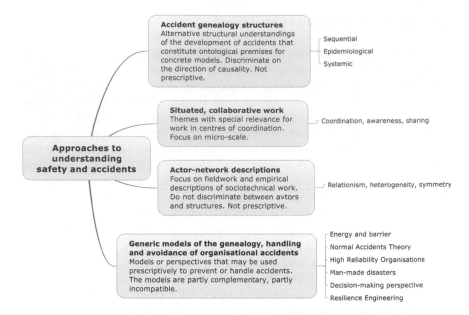

Figure 6.1   Four approaches to understanding safety and accidents. Number four represents the approach used in this study

Chapter 7
Integrated Drilling Operations as Sociotechnical Systems

We have seen how different safety science approaches place emphasis on different aspects of the dynamics of sociotechnical systems and the work within these systems, and how the repertoire may be extended by the ontological perspectives central in science and technology studies (Bijker et al., 1987; Callon, 1986; Latour, 1987; Law, 1986) in general and Latour (1987, 1993, 1999b, 2005) especially. By addressing two cases from the offshore drilling industry and by communicating with some of the more traditional perspectives in safety science the value of this extended repertoire, and how it may affect our understanding of Integrated Operations, will be clearer.

The discussion of how components of sociotechnical systems and the relations between them can be understood is a part of a larger philosophical discussion on how the world is constituted. Being significant for philosophical inquiry should not, however, make us think that it has no practical relevance for the very physical and empirical world of Integrated Operations, and the way we understand and evaluate the safety of these operations.

Integrated Operations may be seen as a special case of the general category of sociotechnical systems, and sociotechnical systems are addressed by much contemporary organisational safety research and literature. Different approaches emphasise different aspects of such systems. In Normal Accident Theory (NAT) (Perrow, 1984; Sagan, 1993) the focus is on the mismatch between complex and tightly coupled technostructures and their human operators; in the literature of High Reliability Organisations (HRO) (La Porte, 1996; La Porte and Consolini, 1991; Weick, 1987; Weick and Roberts, 1993; Weick and Sutcliffe, 2007) the focus is on the qualities of organisations where a complex technostructure is successfully controlled; Reason's (1997) approach to organisational accidents demonstrates how management factors, procedural, technical and cultural factors produce the system safety in concert.

Although these approaches in many respects differ from each other in the way they include and emphasise different elements and characteristics of sociotechnical systems, they all treat sociotechnical systems as complex entities made up of factors belonging to the well-defined realms of humans, technologies and organisations. Few contemporary researchers repudiate the holistic ambition of organisational safety approaches, although there is a great variation in the way it is translated into different models of safety and accidents and in the weight put on the different factors.

Without disagreeing at all with such a holistic perspective, I suggest that there is a fundamental aspect of sociotechnical systems that is not treated explicitly by these organisational approaches. This aspect involves two themes: the first is the way the components influence each other by constantly shaping and changing each other, and what this means for the relational understanding of the dynamics of sociotechnical systems. The second is the problematic demarcation of the system components themselves. Here it is shown that the relevant division of systems into different components or factors is not given in advance and does not necessarily follow the intuitive dividing lines between humans and technologies or between different technological components. These two themes are related to central themes in the sociotechnical writings of Latour (1993, 1999b), whose ontological considerations support the argumentation in this book.

In this chapter we will elaborate on drilling operations as sociotechnical systems by investigating two cases from the offshore drilling industry. This industry is gradually changing from a work process regime marked by a pronounced division of labour and a lack of cooperation across domains to a regime where humans, technologies and the flow of information are closely integrated and where decisions are made by a collective rather than by individual actors. This regime of Integrated Operations is a good case to elaborate on the characteristics of modern sociotechnical systems because IO explicitly addresses the problems associated with treating systems as a collection of demarcated components.

7.1 From Traditional Offshore Operations to Integrated Operations

Integrated Operations denote the vision of a current change process in the petroleum industry. Offshore drilling operations represent a domain where IO are expected to play a major role with respect to speed, quality and safety of operations (Løwén et al., 2009; OLF, 2005; Ringstad and Andersen, 2006). New technologies and new work processes are introduced to get a closer integration of the different disciplines and the data and information that circulate between them. It is this integration that is associated with the *faster, better, safer* maxim (Ringstad and Andersen, 2007).

The *technologies* introduced as IO technologies include real-time updatable geomodels and reservoir models, wired drill pipe, automated diagnosing systems, extended use of sensors along the drill string, automated drilling systems, down-hole seismics and video conferencing systems. The focus on *work processes* in Integrated Operations is largely on collaborative work. Team decisions are to replace individual decisions, and the real-time cooperation between disciplines and geographical locations should presumably be improved (OLF, 2005).

Work processes and technologies are intertwined in a technology-dominated offshore industry. Automated diagnosing systems will not work without evaluating and choosing between the different scenarios proposed by the system or without proper ways of collecting operational experience, feeding it back into the system and thus building its intellectual capacity (Iversen et al., 2006; Rommetveit et al., 2008b; Shokouhi et al., 2009). The different and geographically distributed disciplines will not be able to make team decisions without being audiovisually connected through screens allowing for real-time sharing of data and information that are collected, processed and presented by means of technology (Grøtan et al., 2009). It is this heterogeneous mix of actors, so characteristic of Integrated Operations, that has been studied and analysed to elaborate on the relations and components of sociotechnical systems.

7.2 Method of Data Collection and Analysis

The cases are analysed using a combination of several methods. The main body of empirical data stems from reviews of technical papers and research papers on IO technologies and work processes, on existing reports on an accident on the Norwegian continental shelf (Brattbakk et al., 2005; Schiefloe and Vikland, 2009; Schiefloe et al., 2005) and an existing meta-analysis of the same accident (Antonsen, 2009a). In addition, participative and observation studies, discussions and interviews have informed the elaboration on three specific innovations in decision support and automation for drilling operations.

Participation in different fora where new technologies and work processes have been discussed has contributed to insight into these innovations as well as the dominating discourses associated with them. Regular meetings in the research programme for drilling and well construction and the programme for new work processes and enabling technologies at the IO Center[1] have also been important arenas for participant observation as well as theory development. Conferences arranged by the centre, where safety management, industry and researchers within the fields of petroleum, information technology and safety science have discussed Integrated Operations across traditional discipline borders, have also been important events in which the research has been informed.

Interviews and literature review on the specific technologies gave the author a deeper understanding of the sociotechnical composition and functioning of the innovations. Thirty-five onshore-based informants have been interviewed on the role of these innovations in future operations. The informants include employees in different operating companies, developers in independent IO technology development companies, people from service companies, employees in an operating company's central support centre, researchers on drilling technology and well stability and members and leaders of rig teams. Although the informants' work is mainly onshore, most of them have offshore work experience. The literature study is based on journal articles and internal reports where the technologies are

---

1   The Center for Integrated Operations in the Petroleum Industry conducts research, innovation and education within the IO field. The centre was established by different research and education institutes in collaboration with major international oil companies and suppliers (for more information, see www.ntnu.no/iocenter).

described and the challenges connected with their implementation and use are elaborated.

The argumentation is based on the reviews of a case and a scenario. The case is based on the accident on the offshore oil platform Snorre A on the Norwegian continental shelf in 2004. During preparations for a sidetrack,[2] central well barriers were removed and other barriers failed, leading to a free flow of gas from the reservoir to the surface underneath the platform. The Norwegian Petroleum Safety Authority characterised the accident as one of the most serious ever in Norway, with a catastrophic potential. Unfavourable weather conditions could have had wide-ranging consequences: formation of a gas cloud on the platform; ignition of the gas cloud; a persistent fire; escalation of fire to the riser; difficult and risky evacuation conditions; loss of lives; weakening and potential loss of the platform; damages on the template with 42 wellheads on the seabed; extensive environmental consequences from the leakage of oil and gas (Brattbakk et al., 2005). The current blowout in the Gulf of Mexico resulting from the 20 April 2010 Deepwater Horizon drilling rig explosion is thus not an irrelevant, worst-case scenario. The inquiries showed that the accident emerged and was only just prevented due to an historical co-development of technological, human and organisational resources. The technical standard of the platform had deteriorated throughout the years as a consequence of low priority of investment and maintenance, due to organisational turbulence and a difficult economic situation. On the other side, these organisational and technological conditions, in addition to the platform's historical limited interaction with its surroundings, had affected the crew members and contributed positively to their internal trust, their focus on collective safety rather than personal risk, and their competence and improvisation capability (Schiefloe and Vikland, 2009). This case deals with the first theme, namely the significance of the relations between components of a sociotechnical system in shaping the properties of the components.

The *scenario* deals with the chapter's second theme – the problem of clearly defining the boundaries of an actor. The scenario elaborates on a group of offshore technologies that are currently under development and thus have not been adopted in operations except for some pilot studies. Although they are stand-alone tools being developed separately and partly in competition with each other, once they are adopted their demarcation is shown to be less clear; the way each of these tools is used, modified and combined with other tools overrides the properties of the single tool.

Although the two cases topically seem very different, they are just as connected as the two points they illustrate. That sociotechnical systems are characterised by *relations between* rather than by the *properties of* the actors, and that the demarcation and the properties of actors themselves are context-dependent, can be viewed as arguments for the same phenomenon at different system levels. While the first case shows the relevance of the *relation* between the analytical categories of humans

---

2   A secondary well bore drilled away from the original hole.

and technology, the second case shows that the categories themselves also depend on relations; the properties of different, seemingly independent technologies are highly influenced and formed by the relation to other technologies, as well as their human operators. In a future where drilling operations are more integrated, technologies similar to those in the scenario will play a more important role in situations such as the one outlined in the case.

7.3 Science and Technology Studies in Relation to HRO and NAT

In the elaboration of the themes through the analysis of the case and the scenario, the sociotechnical perspective of existing organisational and systemic theories of safety will be contrasted with dominating sociotechnical perspectives of science and technology studies.

NAT and HRO represent two highly influential approaches to organisational safety, or safety of sociotechnical systems, and over recent decades they have been subject to controversy on the possibility of controlling tightly coupled, highly complex systems. NAT (Perrow, 1984; Sagan, 1993) and HRO (La Porte, 1996; La Porte and Consolini, 1991; Weick, 1987; Weick and Roberts, 1993) come to different conclusions with respect to the safety of such systems. Whereas NAT claims that accidents in such systems are impossible to avoid, and hence *normal*, HRO is more optimistic with respect to what can be achieved by the people and the organisational arrangement of the systems. Common to these two approaches, however, is their treatment of the social and the technological dimension as clearly defined parts of a system. The difference between them lies in whether or not the social sphere is able to control the complexity of the technical sphere.

This is a controversy that has been tackled by elaborating on and drawing in parallel from both perspectives (e.g. Snook, 2000) and by arguing for the cross-fertilisation of the two perspectives (Rijpma, 1997, 2003). Systemic theories (Hollnagel et al., 2006; Leveson et al., 2009; Marais et al., 2004) represent another approach that seeks to develop the organisational safety theory by accounting more thoroughly for the complexity of sociotechnical systems. These contributions underscore that a holistic, systemic approach must address the relations, not only the components:

> Accident models that consider the entire socio-technical system must treat the system as a whole, taking into account all facets relating the social to the technical aspects, and not just the parts taken separately. They must consider the relationships between the parts of systems: how they interact and fit together. (Leveson, 2004: 249)

The systemic approach addresses an important characteristic of sociotechnical complexity and differs explicitly from 'HRO and standard engineering approaches' by focusing on 'the integrated socio-technical system as a whole and the relationships between the technical, organisational, and social aspects' (Marais et al., 2004: 11). Still, the systems approach does not challenge the fundamental epistemology of the approaches it criticises. The transfer of focus from the separate parts to the relations between them is advocated without questioning the nature of the parts and the nature of the relations – or the ontology of the system. Thus, the desired focus on the relations is still constrained by the view on relations as something found *between* the technical, organisational and social aspects. In other words, the *relata* still comes before the *relations*.

Sociotechnical systems have been subject to much research within the field of science and technology studies (Bijker et al., 1987; Callon, 1986; Latour, 1987, 1988, 1993, 2005; Law, 1986). This research has demonstrated the power of explaining sociotechnical systems by their *relations* rather than by their actors (or components), and it has shown how the properties of the actors are *results* of the relations, not vice versa. Latour consequently does not operate with groups of humans and groups of technologies as basic categories and the relations between them as generic relations. The point of departure is a heterogeneous mix of humans and non-humans, and the relations represent occasions of interaction that contributes to shaping the actors. It is important to note that the relations are not seen as channels through which actors pass on their properties. The relation that Latour speaks of, and which he calls a translation, is 'a relation that does not transport causality but induces two mediators[3] into coexisting' (Latour, 2005: 108).

These characteristics make sociotechnical systems difficult to represent by generic models with predefined actors and relations. The advice of Latour is to produce descriptions of the actions that actually take place within the sociotechnical system without referring to general factors or groups such as technology or organisation as points of departure, since 'there is no group, only group formation' (Latour, 2005: 27). The relevant relations and actors should be identified in each case. It is by describing the relations and the actors interacting in a sociotechnical system that one can understand how and why the system dynamics appear and what their consequences, such as safe or unsafe operations, are really made of.

In these studies, the a priori properties of technical and social actors are not seen as determinants for the outcome of events. Rather, those traits are *results* of the rich interplay between actors who themselves derive their competences from this interplay. This is underscored in Latour's elaboration of Pasteur's heterogeneous work of discovering the yeast:

> No event can be accounted for *before* its conclusion, *before* Pasteur launched his experiment, *before* the yeast started to trigger the fermentation, *before* the meeting of the Academy. If such a list were made, the actors on it would not be endowed with the competence that they will *acquire* in the event ... This list of inputs does not have to be completed by drawing upon any stock of resources,

---

3   Latour uses the word mediator for an actor to distinguish it from entities that participate in events without making a difference to the change induced by the event.

since the stock drawn upon before the experimental event is not the same as the one drawn upon *after* it. (Latour, 1999b: 126)

In summary, relations could be understood as *events that change the participating actors*. Furthermore, a clear dividing line between the actors, or the components of the sociotechnical system, such as the technical, organisational and social aspects, cannot easily be drawn since in practice those dividing lines are blurred. Also social anthropologist and cybernetician Bateson (2000: 153–154) subscribes to an ontology where 'the *relations* are to be thought of as somehow primary, the *relata* as secondary', implying that the parts are defined 'solely by their relationships'.

The significance of relations for understanding and controlling complex sociotechnical systems is also acknowledged by other approaches. However, there are some fundamental, ontological differences between these approaches and the approach suggested in here. As already mentioned, the systemic approaches (e.g. Leveson et al., 2009; Marais et al., 2004) do not challenge the established order of the relations as secondary to the components. Further, while the HRO approach is concerned with collective mental processes and the *social* interrelations[4] contributing to those processes (Weick and Roberts, 1993), the lack of reference to non-social dimensions when describing the processes that create order in the chaotic sociotechnical system of an aircraft carrier is striking:

> First, people on carriers are preoccupied with failure ... Second, people on carriers are reluctant to simplify ... Third, people on carriers maintain continuous sensitivity to operations ... Fourth, people on carriers have a commitment to resilience ... Fifth, people on carriers maintain deference to expertise. (Weick and Sutcliffe, 2007: 37–39)

Instead of unilaterally describing the collective social properties and *people on carriers* as driving forces of the system, both the social properties and the system outputs are understood as *results* of the relations between a heterogeneous mix of humans and technologies. This will be illustrated by the case and the scenario presented in the next chapters. The elaboration of the two themes referred to in the introduction results in two points, which are considered important characteristics of sociotechnical systems and hence important findings in the study.

7.4 First Point: Components are Shaped by Relations (Case)

This point is illustrated by an accident that occurred on the oil platform Snorre A on the Norwegian continental shelf in 2004 (Brattbakk et al., 2005; Schiefloe

---

4   In addition to social processes and practices, HROs have been characterised by their unique structural features (Weick and Sutcliffe, 2007).

and Vikland, 2009; Schiefloe et al., 2005). In connection with preparations for a sidetrack from well P-31A, gas started flowing into the well and up in the sea underneath the platform. Not until the day after was the situation under control.

In technical terms, the cause of the accident was easy to establish; during the pulling of a scab-liner (steel tubing) to prepare for the sidetrack operation, central well barriers were deliberately removed; a cement plug was pulled (and the 9 5/8" casing that had been damaged many years earlier was exposed) and the blowout preventer (BOP) was partially deactivated. The pulling of the scab-liner induced an under-pressure in the well and an influx of gas from the reservoir. The primary well barrier, the drilling fluid, was thus mixed with gas, resulting in reduced specific gravity. The outer 13 3/8" casing could not withstand the pressure from the gas, and when this casing broke, the gas flowed freely to the surface (see Figure 7.1). However, as the investigations revealed (Brattbakk et al., 2005; Schiefloe and Vikland, 2009; Schiefloe et al., 2005), the cause of the accident was not to be found in the lack of barriers alone.

Just as with the *Challenger* accident where the O-ring played a central part in the accident (Vaughan, 1996), there existed an underlying organisational context that allowed the technical decision to take place. However, the point here is not the importance of a holistic, organisational approach. The point is that taking the different system components, such as the technical and the social domain, as the starting point for understanding the system may lead us astray because these demarcated components do not necessarily constitute a relevant classification of the system. As Latour (2005) and Bateson (2000) point out, and as the case will show, the relations, understood as the occasions for *constituting and changing* the components, come before the components and thus deserve an attention that is not predefined by the system's components and their properties.

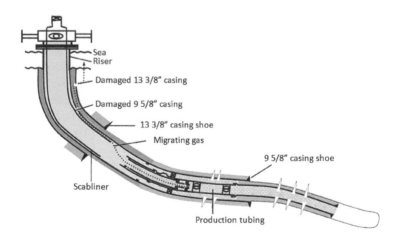

Figure 7.1   The subsurface situation during the blow-out

Source: adapted from (Schiefloe et al., 2005)

7.4.1 Sociotechnical Relations in the Snorre A Accident

Two works connected to the Snorre accident illustrate that explaining the accident in terms of pure social, pure technical or a mere *combination* of social and technical factors without taking into account the relational aspects that are important for shaping these factors involves a risk of reductionism.

The first work (Antonsen, 2009a) is a meta-analysis of two studies of the safety culture at the Snorre A platform performed before and after the accident illustrates the limited explanatory power of an isolated social factor in a sociotechnical system. A year before the accident happened, a survey of the safety culture had been conducted on the rig (Kongsvik, 2003). The safety culture described by this survey differed dramatically from the description in the investigation undertaken after the accident (Schiefloe and Vikland, 2009; Schiefloe et al., 2005). The first survey did not offer any warnings that something serious was brewing. Yet, a year after the accident, the more qualitative investigation revealed that the safety culture had not been satisfactory at all.

There are several possible interpretations of the discrepancy between these two accounts. Antonsen's (2009a) interpretation is that safety culture surveys have little predictive value and that they will naturally differ from more holistic studies with a broader analytical scope. Another obvious interpretation is that the comparison of the studies is asymmetric in light of the hindsight is 20/20 argument. While there is no reason to doubt the relevance of these interpretations, the accident and the studies offer an additional insight which these interpretations do not address; it is not insignificant whether a system is described by its components or its relations. While the first study that focused largely on a cultural component showed a lack of relevance to the unfolding of events, the second study was more oriented towards relations, a methodology that in retrospect appeared a more realistic system description.[5] Thus, the assessments of organisational culture undertaken before and after the accident did not account for the same culture because the former investigation did not take into consideration the relational aspects that played such an important role in the latter.

The second work is a post hoc account (Schiefloe and Vikland, 2009) of the organisational context leading to the accident; the platform's technical condition was shown to have deteriorated due to 'the platform's turbulent history, and the problematic economic situation of the first operator of the field' (Schiefloe and Vikland, 2009: 11). The technical condition was transformed by the organisational rearrangement and the economic situation. Concurrently, the deteriorating technical condition of the platform played a role in shaping the competences of the rig crew: the technical condition and the working situation 'resulted in a well-

---

5   The comparison between the two studies can for obvious reasons never escape from the hindsight is 20/20 argument. While this argument is relevant and important, it does not address the main point that is highlighted here, namely the difference between descriptions based on components and relations.

developed ability to be versatile and improvise. The Snorre crew could handle almost any technical challenge in a swift and competent way' (Schiefloe and Vikland, 2009: 10–11).

The successful emergency handling is attributed to capacities growing out of these relations. One side of this is how knowledge of local technology rendered possible improvisation, such as the recoupling of the electricity supply as well as the cement pumps in order to kill the well by bullheading.[6] Another side is the persistence and mutual trust that had developed through the years:

> The 35 remaining men worked all through the night, under extremely demanding conditions and facing a considerable personal risk. Illustrative is the fact that they considered full evacuation three times during the night, each time deciding to keep on fighting ... [N]obody were forced to stay, but they chose to remain to fight the blow-out, even if they had to work in survival suits and continually be prepared for a run to the life boats. The story told afterwards concerns factors as trust, companionship and identification, all of them elements which can be characterized as aspects of local, social capital. The handling of the blow-out was also characterized by well developed ability to cooperate and improvise. (Schiefloe and Vikland, 2009: 3–5)

At this stage, the roles of technology and humans were mixed to such a degree that it no longer made sense to talk of a technological condition as a property demarcated from human capacities. Neither the improvising capacity due to the technological history of the platform, the competence rendering possible the unorthodox configuration of the power supply and the cement pumps, nor the team spirit and mutual trust due to the history of organisational isolation (Brattbakk et al., 2005; Schiefloe and Vikland, 2009) is adequately explained by referring to static social or technical conditions.

The investigation revealed that 'some of the specific organizational attributes which explains why the incident could happen, were the same which later made it possible to prevent a catastrophe' (Schiefloe and Vikland, 2009: 1). This need not be as paradoxical as it seems. On the contrary, it supports this chapter's first point; that the organisational/social attributes that were found to cause the accident were the same as those central in the successful emergency handling (Schiefloe and Vikland, 2009; Schiefloe et al., 2005) is only paradoxical when organisational/social attributes are analysed as factors demarcated from the technological condition. However, Schiefloe and Vikland points to the *relations* between and the *shaping* of the actors in the history of Snorre A:

> Some of the high-ranking officers told that the investments for a long time had been kept to a minimum, resulting in an incident-driven mode of working.

---

6   Forced pumping of fluids into a formation to force inflowing well bore fluids back into the reservoir.

'Continuous fire-fighting'was an expression used to describe the daily working situation for the technical crews. This was said to result in insufficient opportunities to work in a systematic, planned and long-term way. At the same time, however, these experiences resulted in a well-developed ability to be versatile and improvise. The Snorre crew could handle almost any technical challenge in a swift and competent way. This kind of experience and competence came to play a decisive role in handling the crisis situation when the blow-out occurred on 28.11, when the ability to improvise and find new and innovative technological solutions came out as vital. (Schiefloe and Vikland, 2009: 11)

The technical condition of the platform cannot be seen in isolation from the investments or from the crew and their work practices. Nor can the competences of the crew be seen in isolation from the platform's technical condition. Furthermore, without studying the relations between the components of the sociotechnical systems, there is no way to foresee what difference an improvising crew would make to the system: would it allow the crew to break the rule of always having two barriers against the reservoir or not? Would the crew be able to handle the technical equipment sufficiently creatively to avoid a catastrophe or not? These are questions that must be answered based on the way actors are shaped through interaction and time.

The story of the Snorre A accident can thus be read as a case of Bateson's (2000) argument of the primary status of relations and the secondary status of those entities that are related. The dynamics between the social and technical resources on the platform played an active part as both a cause of the accident and of the successful emergency handling. If this is in any way paradoxical, it is not due to the *course of events*, but to their asymmetric *descriptions*. Holistic explanations of sociotechnical systems are about more than taking into account the different components of the system. Unless sociotechnical systems are understood by understanding their *relational* character, properties of predefined, demarcated components will at best offer ambiguous indications of the system safety.

7.5 Second Point: The Boundlessness of Technologies

This section treats a group of technologies that are developed as a part of the offshore industry's IO strategy. After a brief introduction of the innovations, the second point of the chapter will be elaborated with the help of a scenario showing the difficulties of unambiguously drawing divisional lines that demarcate categories for analytical purposes in sociotechnically integrated systems.

7.5.1 Tools for Interpretation and Automation

The three innovations that are explored are expected by future potential users to affect both onshore and offshore management of drilling operations. The innovations differ in purpose and functionality, but all are designed to handle drilling data in a more effective manner than today, giving way to visualisation, automation and decision support through diagnostics and prediction. To make references to the tools easy whilst retaining their anonymity, they will be referred to as Alpha-Drill, Beta-Drill and Gamma-Drill.

Alpha-Drill is a system that is based on mathematical models of the drill string, the well and the hydraulic conditions in the well. During drilling, the tool is fed with drilling data so as to calibrate and update the models in real time. Thus, Alpha-Drill offers a continuously updated theoretical model of the state of the well, calculates how this model should change with any changing parameter, and compares real-time measurements of the well properties with the expected, theoretical model so as to discover any discrepancies and give warnings before any deviances exceed acceptable limits and the situation escalates. One example could be a situation when the pump rate is intentionally increased but the standpipe pressure stays unchanged. The theoretical model of the system would, in such a case, expect an increase in the standpipe pressure, give a warning signal and if possible suggest one or a set of alternative explanations for the discrepancies.

While this feature of Alpha-Drill is largely black-boxed (Latour, 1987), the tool's other feature produces transparency by means of a representation/ visualisation module. Here, the results from the calculation module present the drilling operation in a well-arranged manner so that the users can quickly be given an overview of the operation. In this way Alpha-Drill also facilitates cooperation and communication within and across disciplines. Alpha-Drill should thus not be regarded as a delimited, clearly defined part of a larger mosaic-like system. It should be regarded as a tool made up of the relations between a range of entities; take away the well properties, the drilling data and the operators negotiating and making decisions based on the representations on Alpha-Drill's graphical user interface – and the tool is an empty box that makes no difference.

Beta-Drill is a tool that resembles Alpha-Drill in functionality, but differs in the underlying logics. Similar to Alpha-Drill, Beta-Drill has predictive functionalities and may warn when measured values in the drilling process differ from expected values. The prediction, however, is based on patterns constructed from experience values and not theoretical descriptions of the mechanisms in the well during drilling. An example could serve to illustrate how it works: during drilling in a certain location, one has collected and systematised experience data for a range of drilling parameters, such as rate of penetration, equivalent circulating density, weight on hook and pump rate, all tied to depth and time measurements from previous wells in the same area. The patterns of such data can serve as *fingerprints* of the drilling operations. During drilling of a new well in the same area, one would expect a minimum of resemblance between the fingerprints of the new well and the 'old' wells. The fingerprint of the actual well will therefore continuously be compared to the historical fingerprints, a process which can produce a corresponding deviance from the pattern of the previous wells whose corresponding problems and successes are known. Whenever the deviance exceeds a certain value, an alarm will be triggered. Another option is to compare the current fingerprint explicitly to fingerprints of previous mishaps. In that way, previous mishaps will appear on the screen like objects on a radar, the nearness to the bull's eye indicating the similarity and proximity of the current fingerprint to that of the mishap. The tool is based upon the theory and logics of case-based reasoning, which is a generic method used to solve new problems based on recognition and solution of similar problems in the past (Shokouhi et al., 2009).

Beta-Drill can be used by the driller, who is physically drilling the well to evaluate the process on a minute-to-minute basis. Those not directly engaged in the drilling operations, such as managers and people from adjacent disciplines who want to keep an eye on the progress and status, have less in-depth knowledge of the specific operations and are therefore not in a position to interpret the information from the tool in the same way as the drillers. To them, the tool might be a quick and easy way to get an overview over possible negative events and the likelihood of their occurrence. As such, it might facilitate circulation of representations to disciplines and people higher up in the organisation without the need for time-consuming meetings. Beta-Drill might thus also affect the communication across disciplines and organisational borders.

Whether it is used at the sharp or the blunt end, history and experience are mobilised and made relevant to actual operations. This mobilisation is made possible through considerations and negotiations at an earlier stage, i.e. during drilling of previous wells. To be able to compare current situations to previous operations and events, those previous operations must be recorded in a way that makes them detectable and accessible. This is a task that needs to be done continuously; gathering drilling data and drilling events that may be of future value, describing the unfolding of the events, how they were approached and what the result was. In other words, the history needs to be described and categorised to be mobilised and to have a future value. This work is a constituent of the tool itself, in the sense that it defines its properties and its functionality. The technology of Beta-Drill is, in other words, constituted by a mix of hardware, software, well history, interpretations and decisions and the border between the tool and its context is indeed blurred.

The third tool, *Gamma-Drill*, is a device for automating different sub-operations of the drilling process that are especially prone to human error. The drillers are responsible for the drilling crew and their activities on the drill floor where heavy, moving equipment constitutes a constant risk. They are, with a joystick, steering a drill string with a length of thousands of metres and a weight of more than a hundred tonnes, and they are responsible for manoeuvring it within limits less than a metre, in concert with the mud pumps whose pump rate must be accurately adjusted to the movement of the drill string. They cooperate with service companies who deliver services that the drillers depend upon; surveys, directional drilling, mud engineering etc. The driller has several screens interfacing with these different actors; video from the drill floor and a range of drilling data from the different service companies. Relations to, and responsibility for, all these actors implies an almost unacceptable workload for the driller (Ptil, 2007). Gamma-Drill is a non-human delegate (Latour, 1992) to which the driller can hand over some of the tasks that are easy to standardise yet difficult for humans to perform very accurately.

One example of this work whose routines and essential accuracy is delegated to Gamma-Drill is the repetitive task of tripping into the well, which means to run the drill string into an open hole, for example after having changed the drill bit. For this operation, there exist clear guidelines with respect to the tripping speed. When the drill string replaces the drilling fluid that fills the well, the mud has basically two ways to escape; out of the hole and up to the surface, or into the formation through cracks that might be created by the pressure exerted by the mud, caused by running the drill string too fast into the well. To ensure that the mud does not enter the formation, the tripping speed should not exceed a certain threshold value. In addition, placing the bit on the bottom is an operation that requires control over the depth of the hole and, at any time, control over the length of the pipe in the hole. Today this is done in such a way that it is possible to make mistakes with the tally and forget to count e.g. a joint or a stand, which is actually not a very rare event. This could have the effect that the bit hits the bottom at high speed, causing severe damage to equipment, hours or days of non-productive time and, consequently, high extra costs. Other sub-operations that may be automated are the controlled start-up of pumps and placing the bit gently on the bottom after each connection.

The introduction of Gamma-Drill into the drilling operations implies a redistribution of competences between human and non-human actors (Latour, 1992). What Gamma-Drill really does is illustrated by the technical and social work it takes over. In other words, what has to be done today, without Gamma-Drill: the driller and his crew must record in the tally each of the pipes that are to make up the drill string and keep track of every pipe that goes into the well; the driller must perhaps get a roughneck to measure the length of an element of the string to double-check in case of doubt on whether this is really the same element as described in the tally; he must regulate the speed of the pipe so that the operations go fast but still do not exceed the maximum speed limit; he must carefully regulate the start-up of the mud pumps after every connection with the same balancing between efficiency and safety as is the case for the pipe running speed, and when the bit approaches the bottom of the well he must set it carefully down so that no damage to the equipment occurs. In addition, he must answer telephone calls from the concerned data engineer who informs him that the occasional transgression of the speed limit could jeopardise the well's integrity; he must do a few exercises to shake loose muscles that stiffen because of the repetitive movements; he must constantly concentrate, keeping his eyes on the drill floor and the screens, and move his joystick according to his observations.

The delegation, however, of some of the tasks from the driller and his crew to Gamma-Drill does not mean that all work is transferred from a social sphere to a technical sphere so that the methods by which the work performance is evaluated should be changed from socially-oriented to technically-oriented. What happens is that some tasks previously performed by humans are redistributed to non-humans so that social and technical actors become even more interwoven. The performance of Gamma-Drill will depend on new work processes of linking information about the down-hole conditions to the drilling equipment specifications and the relevant threshold for the drilling operations. How to set the threshold values of the automated drilling process, how to quality check the detailed information about every piece of equipment that is run into the well and when to overrule the automation; these questions reflect new dividing lines between a new set of actors involved in the drilling process.

Drawing the line between the social and the technical thus becomes even more difficult than before. However, there is no reason to draw such a line, if it is possible at all, because such a line appears to be a *result* of a system's performance, not an a priori distinction around which it makes sense to organise the description. The exploration of the consequences of the delegation of work to Gamma-Drill illustrates the problem of drawing dividing lines between the actors, be they technical or human.

7.5.2 Scenario: Understanding a Mud Kick During Drilling

The following scenario might illustrate the seamless nature of a sociotechnical system, represented by drilling operations in a future where Alpha-Drill, Beta-Drill and Gamma-Drill are adopted. It describes an important category of events in drilling operations in the sense that it is a situation that occurs frequently and whose accident potential is high.

Such is the scenario: during tripping out of the hole using the automated function of Gamma-Drill, the data engineer gets an alarm from Alpha-Drill indicating mud gains, which could be the initial stage of a blowout. Beta-Drill so far shows a data pattern similar to previous trips, and in those previous operations there are no indications of mud gains. Now, numerous actions can come into consideration to cope with the situation; switching from automatic to manual tripping control by disabling Gamma-Drill; phoning the derrick-man and asking him to manually check the level of mud in the trip tank to make sure that the registration is not due to a sensor error; performing a flow check;[7] consulting Alpha-Drill for support in diagnosis and possible solutions. Further immediate actions would depend on the answers these and other investigations give. However, when the situation is under control, a lot of sociotechnical work will still remain to make sure that the event and the new information is taken into account by all relevant actors. If the irregularity indicated by Alpha-Drill is real, it will be an important experience to feed into Beta-Drill. A report will have to be written about the causes of the event, which measures were taken and what the results of these measures were, and this

---

7 A flow check means to monitor the level of mud in the trip tank while the pipe is held still and the mud pumps are off – in such a situation the volume in the trip tank should stay unaltered.

report should be linked to the depth the event occurred at and all other relevant drilling data at the time. This paperwork is a part of the continuous engineering of Beta-Drill, making it able to detect similar situations at an earlier stage in e.g. the next well to be drilled in the same area.

The use of Gamma-Drill might have to be reconsidered for the remaining tripping operations, since its thresholds showed it to be inadequate in the specific context. A question that would have to be asked is whether this incident should be considered a unique case, not general enough to affect the tool's functionality, or if Gamma-Drill should be adjusted to function more adequate in similar future instances. In that case, this will generate work for the tool provider onshore, perhaps weeks of meetings, reprogramming, testing and report writing to adjust the tool.

The scenario illustrates how not only the independence, but also the demarcation, of the three presumptively independent innovations, and the role of each of them in the network of human and non-human actors, are shaped by other actors in such a way that the properties cannot easily be traced back to each one of them. Instead, they must be acknowledged as properties belonging to the relations *between* them.

7.6 Summing Up: Relations and Boundaries

In descriptions of sociotechnical systems and in explanations of accidents, the primary focus of NAT is the properties of the system's technostructure, and the inability of the organisation to follow and account for the interactions. The argument is, in short, that 'certain technologies make major accidents inevitable, no matter how well managed an operation may be' (Hopkins, 2001: 65), an argument Hopkins (2001: 65) calls 'unashamedly technological determinist'. The case in Section 7.4, however, shows that describing the safety of an offshore oil rig by referring solely to the technical structure and condition conceals that the social sphere is an inextricable part of the technostructure and that the technical condition is an ambiguous determinant of system performance.

On the other hand, according to the HRO researchers, the outcome of industrial processes is largely a function of the organisations'

> (1) preoccupation with failures rather than successes, (2) reluctance to simplify interpretations, (3) sensitivity to operations, (4) commitment to resilience and (5) deference to expertise, as exhibited by encouragement of a fluid decision-making system. Together these five processes produce a collective state of *mindfulness*. (Weick and Sutcliffe, 2001: v)

However, the social and organisational character of offshore operations is strongly dominated by the technologies they seek to control. This aspect misses in characterisations of independent social properties when they are detached from the rest of the actors in the sociotechnical system.

One possible solution to this problem of truly accounting for the interplay between the technical and the social/organisational factors could be to integrate NAT and HRO to account thoroughly and symmetrically for both the technical and the social dimensions of sociotechnical systems. However, by sticking to the same ontology of the sociotechnical system as a composite of separated components and not changing the focus from the *factors* to the *relations*, such integration will not address the problem raised here. Rijpma (1997), on the other hand, seeks a cross-fertilisation between the theories by explaining relations instead of traits, by assessing 'the effects of one theory's independent variables on the other theory's dependent variable'. This is done by asking the following questions:

> Which effects do complexity and tight coupling have on the effectiveness of reliability-enhancing strategies discerned by HRT (high reliability theories) and, hence, on overall reliability? How do HROs' reliability-promoting strategies impinge on complexity and tight-coupling, and, hence, on normal-accident proneness? (Rijpma, 1997: 16)

Without stating it explicitly, Rijpma indicates the usefulness of reconsidering the epistemology of the sociotechnical system understanding; the malleable and non-deterministic nature of social and technical actors working in concert implies that a sociotechnical system will not be well understood if the technical and social are regarded and analysed as predefined, separate components with corresponding properties. The fundamental drivers of events are not components and their properties, but *relations*. Still, the study of relations is not a straightforward task. As shown in the case in Section 7.5, the system components are not easily demarcated since they are themselves characterised by relations. This recursive aspect of the relational nature of sociotechnical systems points to the challenge of choosing a relevant method of description for the sociotechnical system. If it is the relations *between* actors that are important, and if actors cannot be described without reference to other actors, how can the relations be precisely identified?

The seemingly recursive paradox can be bypassed if relations are studied not with well-defined actors as a starting point, but, as indicated in Section 7.3, as *events that change the participating actors*. Such an approach is naturally difficult to standardise. Thus, a relational perspective on sociotechnical systems makes the job of describing the systems and projecting their properties at a later stage a comprehensive task. An understanding of sociotechnical systems would have to be supported by empirical investigations of the change processes that characterise each distinct system. The less generic the descriptions of the actors and the relations are, the more adequate the understanding of the system will be.

7.7 The Advantage of a Relational Ontology

Sociotechnical systems *can* be seen as compound systems where the different components are described separately and where the interactions between the components are what make the systems so complex and difficult to control. This is a common perspective for organisational safety approaches as well as systemic approaches to safety.

Alternatively, sociotechnical systems can be seen as primarily *defined by relations*. The actors are not seen as homogenous components with static properties, but are defined and shaped by the relational context in which they occur. The actors, or the components, would thus be seen as products of the relations rather than the starting point for the analysis of these relations.

The difference between these two perspectives could seem subtle. However, the epistemology on which they are based differs fundamentally, and the difference this makes should have significance for how sociotechnical systems are analysed. If the relevant subdivision of the system into components or actors is a *result* of the analysis rather than a starting point, the starting point should be the *relation*. As a consequence, generic descriptions of sociotechnical systems as interactions between predefined entities forestall system analyses and thus risk taking on a tautological character.

Organisational approaches to safety such as NAT or HRO illustrate this; central to the discourse of these two approaches is the question of whether a complex industry may be safe or not, whether an organisation may be highly reliable due to its organisational characteristics or if accidents will be inevitable due to the complex technological structure and the limited human cognitive capacities. What I have suggested is that the stalemate (Rijpma, 1997, 2003) between the two approaches of NAT and HRO might be due to the line drawn between predefined system components whose relation is considered as a secondary property.

The perspective advocated here – a focus on the relations as primary entities, and an acknowledgement of the fuzzy border between actors – does not promise an easier diagnosis of sociotechnical systems as safe or unsafe than those resulting from approaches such as NAT or HRO. On the contrary, the analyses will be more labour-intensive, less generic and less durable. The gain, on the other hand, is analyses of sociotechnical systems that better match the ontological elasticity and the dynamic character that this chapter has shown such systems to have.

We have now come a long way in localising Integrated Operations along axes of content, methodology, safety theory and ontology. It should thus be possible to recapitulate and target the main objective with higher precision. We have learnt that the question of how we can study the effect of Integrated Operations on operational safety should be reformulated and specified, or else it may not be possible to operationalise and to pursue in a meaningful way. If we instead ask how some specific technologies for monitoring, diagnosing and automating offshore drilling operations may have significance for the sociotechnical work, we will be in a much better position to provide a meaningful answer, and to *thereafter generalise in an attempt to answer also the first question.* This last point is important, because we want our efforts to have validity not only in a very narrow sector, but to address the larger question of how Integrated Operations are affecting the safety of the oil and gas industry in general. It is my hope that the reader will have a much clearer conception of what this general question implies in terms of research, operationalisation and sobriety after the reading of this book.

Let us now revisit the cases presented in Sections 2.1 and 2.2, and try to figure out how what we have learnt from this case may address our main question. The approach will be to analyse the cases as a case of *articulation work* and a case of *shared understanding.* In the analysis of both cases we will thus need some patience and be prepared for a couple of detours to account for the theoretical concepts and their connotations before proceeding with the analysis.

Chapter 8

Articulation Work: Revisiting the Case of Mud Losses[1]

How people work is not always apparent, and many qualities of work performance are seldom brought to notice (Star and Strauss, 1999; Suchman, 1995). Much work is performed in such a way that only those who actually perform it know how it is done. The ability to account for this invisible work and the tacit knowledge that accompanies it, and to coordinate and integrate it in a wider context, might have great significance for the successful performance of an organisation. When many different work processes that depend on each other are distributed over many different actors and over large geographical distances, articulation work is needed in order to make the different work processes mesh well together. Articulation work is the 'work that gets things back "on track" in the face of the unexpected, and modifies action to accommodate unanticipated contingencies. *The important thing about articulation work is that it is invisible to rationalized models of work'* (Star and Strauss, 1999: 10; my italics).

It is especially important to be aware of and take into consideration the significance of this articulation work when an organisation is undergoing major changes in terms of technologies and work processes. If not, the organisation might unintentionally be deprived of important qualities constituted by the technologies and work processes that are being replaced.

Within the drilling departments of the petroleum industry, development of software, hardware and work processes is a business domain of high priority. Two distinct goals can be identified. First, more instrumentation, data processing and automation will enhance the models, diagnostics and automated responses in drilling operations. Second, more effective sharing of information will ensure that the right information is accessible to the right people, at the right time, thus improving both individual and team decision-making in the drilling organisation. *Integrated Operations* is a common term for this new way of carrying out work.

By introducing new technologies and work processes, one is aiming at improving the outcome of the drilling operations with respect to efficiency and safety. A commonly used slogan is *better, faster and safer operations*. The resemblance of this slogan to NASA's adoption of the *faster, better, cheaper* (FBC) policy is striking. The only difference is that *safer* has substituted *cheaper*. Following a series of mishaps in space science, the NASA's FBC policy was called

---

1   This chapter contains an analysis of the case presented in Section 2.1. The reader is advised to reread this case before continuing the reading of this analysis.

into question. The acknowledgement of the subsequent investigations was that FBC was more a result of pressure and demands from stakeholders than a policy choice (Woods, 2006). There is obviously also pressure in the petroleum industry towards cost reductions, but the choice of words describing future aims might indicate acknowledgement of safety as a key dimension of success.

This success is supposed to be brought about through new technologies and new work processes. Less attention is paid to the work processes that are modified or replaced when introducing such new technologies. If their significance is disregarded, work processes that are contributing to successful performance may be unwittingly lost in the process of change. Instead of seeing articulation work as a burden that should be eased, it is emphasised rather as an integral *resource* for successful cooperative work.

Automation is a central aspect of Integrated Operations. Although automation is undoubtedly a very powerful strategy in drilling operations and in industrial processes in general, there are some issues that should be reflected upon. First, a precise determination of the system boundaries is vital for automation to be successful (Hollnagel and Woods, 2005). As shown in Chapter 7, the boundaries relevant to specific operations or problems are not always well defined. Thus, the automation might circumvent resources and work processes that are not part of the formal, visible work repertoire, but are nevertheless important in dealing with contingencies. Second, since automation implies closing the loop of input and output, automated systems need to be specified fully in order to function consistently. Even with unambiguous boundaries, many systems will rely on decisions and actions that cannot be fully specified. Automation of work processes that depends on tacit knowledge and trade-offs might therefore undermine the necessary performance variability (Hollnagel, 2009).

The objective of this chapter is to describe and put value on the work involved in *coordinating* different actors, their work and their tacit knowledge. The contribution to the field of safety science is the descriptive and conceptual treatment of the role of articulation work to the successes of normal operations and handling of emerging crises. With such an understanding, new technologies and work processes can build from the strengths of their predecessors rather than simply replacing them.

8.1 Theoretical Framework

In much safety research (Hollnagel et al., 2006; La Porte and Consolini, 1991; Perrow, 1984; Weick and Roberts, 1993), complexity and the way it is handled by the social collective is a major concern. However, the weight that is put on the complexity itself and the coping strategies adopted by the social collective differs. Two traditions that have set the tone for safety research during the last 20 years, and that relate to these issues quite differently, are those of Normal Accidents Theory (NAT) and High Reliability Organisation (HRO).

In NAT (Perrow, 1984), much attention is paid to how the system works with respect to complexity and couplings, postulating that loss of control is normal and inevitable in systems with high interactive complexity and tight couplings. Compared with HRO, NAT is a technologically deterministic approach, emphasising that preconditions for safety are implicated by the technical design of the system.

In the tradition of HRO (La Porte and Consolini, 1991; Weick et al., 1999), much emphasis has been put on mindfulness, safety culture and the ability to reconfigure the organisation in times of crisis as a strategy to meet high demands in complex organisations. In this tradition, more attention is paid to the social aspect of the organisation than to the technical descriptions. As such, HRO applies a more social constructivist view to the organisational reality.

In this study, complexity and the way it is handled by the social collective is in focus, but neither the technical structure nor the social act of construction is given priority with regard to making a difference and affecting the conditions for safety. They are not even treated as a priori categories. In line with much of the work within science and technology studies (e.g. Latour, 1986, 1999a), humans and machines, as well as data plots, procedures and other artefacts, are all treated as potential mediators of a given situation and its unfolding. Neither NAT's focus on technostructure nor HRO's belief in the power of the social is rejected; they are just not taken as *starting* points for the investigation.

Rather, the point of departure is the practical work that is actually done, and the way the actors themselves account for it. The concept of articulation work, coined by Strauss (1985) and elaborated by many others (e.g. Gerson and Star, 1986; Schmidt and Bannon, 1992; Star and Strauss, 1999; Suchman, 1996b), is useful, because it offers a framework to understand what the coordination and integration of distributed work really means. More than mere coordination of tasks, it involves the 'meshing' of numerous tasks and clusters of tasks, the 'meshing' of the efforts of various workers and the units they are organised into and the 'meshing' of actors with their various types of work and tasks. In this articulation work interactional processes, such as negotiating, compromising and lobbying, contribute to establish, keep going and revise the cooperative sociotechnical arrangements of the cooperative work (Strauss, 1985). This work is also referred to as an organisational due process through which a local and temporary closure and stability of the system is achieved, often as a compromise, so that work can go on (Gerson and Star, 1986). This articulation work has generally been treated as an overhead cost in the literature (see e.g. Bannon and Bødker, 1997; Schmidt and Bannon, 1992).

Articulation work should not be confused with the notion of workflow, which is more a description of predefined packages of work that can be planned for in advance and that can be analysed from a desk-top position as a non-problematic part of the broader work logistics.

Hampson (2005) has systematised some different conceptions of articulation work, where the work is characterised as either visible or invisible and routine or non-routine. Since visibility of work depends on the point from which it is observed, there is a touch of relativity in the descriptions. Articulation work as such should, for the same reason, not be treated as an absolute entity, since what is regarded as prescribed, documented primary work from a distance really includes much articulation work, which can be seen only through closer observation.

By using articulation work as a framework for examining empirical data from field studies, some features that can serve as generalisations and ways to understand the alignment of organisational safety are produced. It is done through a pragmatic and ethnographic approach, grounded on a relational epistemology. It takes inspiration from the works of, among others, Latour (2005) and Garfinkel and Rawls (2002).

Accounting for the work that is actually done might seem to be an ordinary, prosaic and non-scientific task, both to those working in the drilling organisation and to the researcher. But as Suchman (1996b) claims, 'the better work is done, the less visible it is to those who benefit from it'. This 'invisibilisation' is an intended effect of job specialisation, in the sense that one is thus able to black-box and not worry about the work of others, while at the same time depending upon it. Many studies have been undertaken to reveal how collaborative work is carried out by a collective of actors in different domains, such as airport ground operations (Suchman, 1993), ship navigation (Hutchins, 1995a) and botanical research (Latour, 1999a). The aim of this study is not to reproduce findings from those works, but to develop the learning from them by showing how problem-solving work in a drilling organisation contributes towards rendering the invisible connections visible, to present the complexity of the operations to the involved parties and to obtain local closure and stability through the due process of articulation work (Gerson and Star, 1986). The key point is that this process, which is argued to foster acknowledgement and understanding of the complexities of drilling operations, is really maintenance of organisational safety.

There is a great potential for understanding safety by examining successful operations. A large body of safety research pays attention mainly to failures and accidents, aiming to constrain performance by means of standardisation, barriers and procedures. This is based on the assumption that accidents happen because humans or technology fail and that risks and accidents can be broken down into linear combinations of failures or malfunctions. Hollnagel (2009) finds such explanatory models inadequate for modern industries that have grown increasingly intractable and, as a consequence, underspecified. He suggests that an underspecified system cannot be controlled by prescribing and constraining performance, and that it depends on a continuous performance variability that meets the demands from a system whose description is not static. Accordingly, safety research should pay attention to, and account for, the features of this performance flexibility involved in the production of safe work.

8.2 Method

The case description is based on a field visit to an onshore expert centre of an international oil company. In connection with this specific case, the author spent two full working days at the centre. Six of the expert centre employees were interviewed. These were the centre manager, two drilling plan leaders, one analysis engineer, one well control discipline representative and one geologist. Three of the interviews were formal, structured and tape recorded, while the other three were less informal and structured, and not recorded (notes were taken). All interviews were undertaken at a time when the case was already in progress and some of them after its conclusion. Some of the informants were also contacted on several occasions during the writing process for information and opinions both on the course of the event, for the details of the author's interpretations and for discussions of the overall conclusions.

The time not spent on interviews was spent on observation. Both general and case-specific work was observed. Activities of special interest were video and telephone conferences between the expert centre and the rig team, the geologists and the offshore rig respectively. In addition, the internal discussions across the room were very informative for developing a rich understanding both of the case and the way of working and cooperating.

In every situation where significant changes in the situational status appeared, such as new understandings, disagreements, settlements, decisions etc., the contributing actors and the way their contributions made a difference with respect to the situational status were identified. Contributions from different professionals and roles were not the only interest; different types of technologies, reports and other tools that made a difference in the work processes were also accounted for.

This way of accounting for the actors and their contributions to the organisational achievements follows from the theoretical approach described above, and is inspired by ethnographic and ethnomethodological approaches (Garfinkel and Rawls, 2002). The participants have thus been considered as most competent in producing accounts of their own work, *informing* the researcher, rather than being subject to scrutiny and interpretation.

The method used also bears resemblance to the approach of grounded theory, which could be said to represent a resource for ethnomethodology (Glaser, 1994; Glaser and Strauss, 1967). By not discriminating a priori between different empirical observations with reference to theoretical orientation, it is possible to develop theory with a hermeneutical treatment of the data, allowing the empirical data to speak for itself, rather than only serving to confirm already established theories. Hence, even the research question has been formulated and has grown out of the field that has been studied, through interaction and communication with the participants and through an emerging understanding of what it really means to drill an offshore well.

It has been a methodical issue to involve informants during the rewriting phase of the work. The findings have been discussed with some of the informants, both to ensure that the actual case descriptions were comprehended and described correctly and that the considerations and conclusions are plausible when viewed from the perspective of the informants' profession. In this work the informants had access to logs from the events so that both observed and non-observed events and details could be double-checked.

The fieldwork that forms the explicit basis for the case would have been of less value if fieldwork in other locations had not been undertaken. The drilling organisation is distributed over several locations where work is performed concurrently and cooperatively; it is therefore of great value to understand the work processes in onshore rig teams and offshore rig locations when the work within an expert centre is investigated. It is thus worth repeating that the author has spent considerable time both in onshore rig teams and on offshore rig locations.

8.3 The Field

The drilling organisation is distributed over several different locations. With the onshore rig team as a hub of information and communication, the structure and interaction can be described schematically as below.

8.3.1 The Drilling Organisation

The company that has been assigned the drilling licence serves as the overall manager and decision-maker of a drilling project; this company is called the *operator*. The representative of the operator on a drilling location, the company man, is the person responsible for operational issues. Most drilling-related work on an oil rig is contracted out to drilling contractors and oil field service companies. The drilling contractor will often own the rig and provide the basic drilling services and staff such as driller, derrick-man and roughnecks. A range of service companies provide specialist services, such as directional drilling, mud logging, drilling fluid services and many more.[2]

The onshore rig team is the organisational unit responsible for writing the drilling programme and following up the drilling process. It consists typically of three drilling engineers and three completion engineers, two drilling/well plan leaders and a leader of the whole group, a drilling superintendent. Apart from this team, the main groups contributing to the drilling process are: (a) the offshore rig crew with representatives from operator, drilling contractor and service companies, (b) onshore representatives from contractor and service companies, and (c) a number of onshore experts available for consultancy both for the offshore

---

2   The derrick-man is in charge of the mud-processing area. Roughnecks are low-ranking members of the drilling crew. The mud logging company's responsibility is primary well control.

crew and the onshore team. A conceptual figure of the drilling organisation is shown in Figure 8.1.

The main communication lines, seen from the perspective of the onshore rig team, are those between the onshore rig team and the offshore rig crew representatives, between the onshore rig team and the onshore representatives from contractor and service companies, and between the onshore rig team and different onshore expert centres. This communication is enabled through wide use of video conferencing, which has become an integrated and technically rather routine resource to the organisations that have been studied.

As the onshore rig team is the unit responsible for the drilling operations, all major decisions and significant changes to the planned procedures need its approval. Through one central web portal relaying all surface and downhole measurements, the team has access to much of the historical and continuously produced, real-time data from the well. This tool enables the drilling engineers to monitor the operations and to compare actual trends to expected trends. Inspired by Suchman's description of the operations room for ground operations at airports,

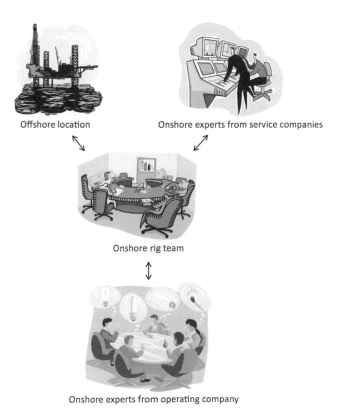

Offshore location   Onshore experts from service companies

Onshore rig team

Onshore experts from operating company

Figure 8.1   The drilling organisation

it seems reasonable to conceptualise the rig team as a '*centre for articulation work across the multiple, distributed settings and interdependent activities*' (Suchman, 1996b).

8.3.2 Administering Drilling Operations

The safety of drilling is about handling equilibriums: the underground formations may contain gases or fluids, and different zones can be under-pressured or over-pressured compared to the pressure of the surrounding formations; while drilling the well, the cuttings need to be transported to the surface by the circulating mud at the same rate as it is drilled; when the drill string is run into and out of the hole, mud can be pressed out of the well or formation fluids can flow into it. Addressing these challenges implies retrieving and interpreting information. What this information really consists of is translations of the properties of the well and the drilling equipment into figures and graphs and numbers inscribed onto papers and screens. Gathering these inscriptions and putting them together into a whole, consistent picture is a large part of what retaining control over the well and responding to events is about. Handling this jigsaw puzzle is made difficult by the fact that some parts may be missing, others may be superfluous and the final result is not defined perfectly in advance because the system – its definition, and its safe operations window – is subject constantly to change.

One of the challenges that today's drilling engineers face, especially those that have little experience, is to make prompt, effective use of the large volume of data generated during the execution of drilling operations. With new sensors and new tools, the amount of data has increased greatly during recent years, and it is reasonable to assume that this development will continue in the future.

The distributed character of the information also poses a challenge. The historically developed division of labour within the drilling organisation has resulted in a division of information. Although all information stemming from the well may be technically available to the rig team, it is not effortless to reconstruct the diverse array of data in such a way as to produce a representative understanding of the well. The data may be gathered geographically, but still be distributed epistemologically. This is how it was expressed by a member of an onshore rig team:

> The problem is that there is so much data, and people have so much to do. And something that is obvious for me is not obvious for, say, the geologist. There was a case here two months ago where a geologist had vital information that we didn't have. It was available, but nobody had told us that there was a gas zone that we shouldn't drill into. And so we drilled into it. So, they had the information, but we didn't get it.

In the following chapter, the process of putting together the relevant information is analysed with respect to the case presented in Section 2.1.

8.4 The Articulation Work of a Mud Loss Situation[3]

The cause of the mud losses was identified after a few hours, having involved a range of actors: the offshore rig crew, the members of the expert centre, a well drilled some weeks ago, drilling plots from the two wells, a new mud system whose properties were not well understood, MS Excel, two geologists in another city and finally, the shale shaker on the rig. Before the mud losses occurred, the rig team needed far fewer resources to handle the drilling process. Using the terminology of Latour (2005), the complexity of the system was low. The sociotechnical system was stable, kept together by a range of invisible actors whose work was more or less black-boxed. The work inside these boxes proceeded smoothly without being accountable to outsiders, matching Suchman's (1995) already-mentioned notion on work rather well: 'the better work is done, the less visible it is to those who benefit from it'. Once the equilibrium broke down, in this case leading to mud losses that could not be accounted for, the black box was opened and the multitude of actors working in the background to keep the well under control, appeared. *Now* the work of drilling became complex.

The conclusive interpretation might seem trivial compared with the three other conjectures. The disclosure of the mud losses over the shakers was made by visual inspection. Had this inspection been made at an earlier stage, the case trajectory might have been different. However, the complexity of drilling operations contributes to a number of contingencies. The handling of this complexity involves a trade-off between efficiency and thoroughness (Hollnagel, 2009). Herein lies the answer to the question why nobody inspected the shaker at an earlier stage: the mud losses were small and a quick visual inspection may very well have been made by the shaker hand[4] at a point when the losses were not noticed. Before the mud logger actually did so, others could also have checked the shakers. The mud engineer could have gone for a look, but he is not under any obligation to do so, because primary well control is not his responsibility. At the time, he was busy supervising a subordinate while they mixed new mud to replace the losses. Although primary work responsibilities may be clear enough, situations will arise that are not clearly confined within the borders of primary work. The handling of such situations thus requires articulation work.

8.4.1 Displaying and Negotiating Complexity through Articulation Work

When drilling proceeds smoothly, the well responds to stimuli as expected. When the mud sticks to the cuttings and is lost over the shakers, it is the onshore rig team's responsibility to organise the gathering of necessary actors in order to make sense

---

3   Cf. the case in Section 2.1. The reader is advised to reread this case before continuing reading the analysis.

4   The shaker hand is a crew member dedicated to watching the shakers. However, he does not do this continuously as this is not his only task.

of, and stabilise, the emerging situation. The well and its constructors go from constituting a sociotechnical system of low complexity to a sociotechnical system of high complexity, and the activity of the actors has to increase accordingly. That is why the expert centre is contacted, why a well of the past again makes a difference, why the geologists must evaluate the relationship between a new type of mud and the geological properties of the well bore formations. Such work of re-establishing control in the face of unanticipated contingencies relies not simply on the primary work defined and demarcated by the division of labour, but also on the situational alignment of the distributed work into a pragmatic configuration that helps solve the problem (Schmidt and Bannon, 1992).

The articulation work affects a range of actors, both human and non-human. Different constellations of actors are emerging, of which some may be unprecedented. What are they doing, these groups of actors that are gathered on the basis of the different conjectures about what is going on? One way of describing their work is to say that they are negotiating between themselves the credibility of the conjectures. The conjecture that the causes of the mud losses were in common with those in the well drilled some weeks previously was accounted for by two of the members of the expert centre and two Excel plots. Another conjecture was tested by a group of two geologists, the mud system and the geological formation. The conjectures are not picked randomly, but are established on the basis of available information, knowledge and experience. Associative thinking seems to play an important role, empowering cognition with propositions based on similar events in the past.

In these group conferences, several black boxes are opened. The negotiations constitute opportunities and arenas for the invisible work and actors to be presented. Before the event with the mud losses, hardly anyone thought of the previous well and the mud losses that were experienced in that job. Specific attention had not been paid to the connection between the new mud system and the formations that were to be drilled through. Actually, the considerations on which the choice of mud was based were not clear. The shale shaker's inability to separate the mud from the cuttings had not been considered or at least was not discussed. The calculations of suitable mud density, which had been done by a service company, had not previously been called into question. Rather than unambiguously representing a weakness in the work process design, its black-boxed character is also a central feature of the division of labour (Suchman, 1995). But suddenly, with the increased complexity due to the mud losses, all these connections were made visible. Some of them did not have explanatory power to account for the mud losses in this specific case, but that did not make them less relevant. What all the reasoning contributed to was to make visible the many aspects of drilling, which dependencies existed and which scenarios might possibly unfold at any time. The black box was not permanently closed.

Through engaging in solving the problem with the mud losses, many issues were brought into play. Central actors in the drilling operations were connected to other less visible actors that were not considered relevant at an earlier stage of the operations. These new assemblages, or groups, could be described as emerging constellations. Their existence is transient. They don't define permanent, structured constellations before they are established to account for the mud losses. They are not planned for in advance and, after the event, they dissolve. They leave traces, though, by the difference they make in solving the problems, and in the memory and knowledge resulting from the experience. Such events should, therefore, be considered as resources to the organisation. They also offer hints to how more prosaic and regular events, such as morning meetings and review of detailed operations procedures, can serve as similar occasions for displaying and negotiating the complexity of the sociotechnical system.

The onshore rig teams are responsible for the success and the safety of the operations, from the start of planning to the end of completion. Still, their experience of the drilling operations appears to be somewhat abstract. To them, drilling is mainly paperwork, meetings and discussions. New drilling engineers are surprised that being a drilling engineer is really being a project leader, coordinating and overseeing other people's work. But even this task verges on abstraction. As one drilling engineer said after receiving applause following the completion of a successful well; 'how did this well come about, really? And what was my contribution to this?' When tracing the contributions by such a central actor *after* the well is completed is difficult, small wonder that it might be difficult to account for the contributions *during*, or even ahead of, the operations. Still, this articulation work seems to be important both in terms of getting the work done, and in terms of creating understanding and knowledge of the complex relationships in drilling operations.

8.4.2 Articulation Work – from Burden to Asset

Articulation work is often considered as an overhead cost, a workload that is complex and demanding. Much effort is thus put into reducing this workload, and enhanced or new ICT systems are often central parts of the solution (Schmidt, 1996). This view on articulation work obscures the fact that the ways in which the meaning and effect of different types of work is experienced is relative to the point of observation. What is seen as articulation work from one point of observation could be seen as primary work from another. Thus, the work of the rig team, which consists of more than simply chaining together packages of primary work performed by other actors, would perhaps benefit from enjoying a status closer to that of the primary work it accompanies.

This is indeed a general feature of much work: as tools for facilitating and easing primary work are introduced, the remaining tasks appear more and more as articulation work. One way of conceptualising this is to imagine a delay between the transition of work to a higher level and the acknowledgement of this work as core, primary work. Turning the articulation work from a burden into an asset is perhaps dependent on redefining it into primary work again. Once defined as primary work, the articulation work might be taken more seriously in development processes and not be regarded simply as inefficient use of resources and time.

8.4.3 Implications for Organisational Change Strategy

The value of articulation work is often not taken into consideration when work processes are discussed. New tools for handling real-time data from the well during drilling through monitoring and decision support, possibly also through automated responses to deviances, could serve as examples of efforts to reduce the need for articulation work. While much effort is made to reduce the need for articulation work by new tools and work processes, findings in this study indicate that articulation work per se has a value that is different from what some of the substituting tools are designed and expected to offer.

When new technologies and work processes are introduced to the drilling industry, these innovations will represent both modification and replacement of existing practices. As has been illustrated in this study, the articulation work involved in solving a distinct problem also plays a role in both individual and organisational learning and maintains and expands the imaginational repertoire and the understanding of complexity and dependencies. As such, articulation work is a potential source of organisational safety.

Following from the theoretical starting point and the empirical findings of this study, the outcome of the ongoing change process is not determined by how the new tools and practices are *designed*, but rather how they are *performed*. This is again dependent on the acknowledgement of the role articulation work plays in the drilling operations. Forestalling the next chapter, the following example can serve to illustrate this: in a morning meeting, a heterogeneous group of experts were debating possible causal relations behind a problem of inconsistent measurements during the previous night's operations. The depth measurements from the 12¼" section did not match the depth measurements for the same points from the 8½" section.[5] Without knowing the reason why the two sets of measurements deviated from each other it was impossible to evaluate the reliability of the future measurements for landing the well in the reservoir in accordance with the plan. Different explanations that could account for the discrepancies were discussed. Among the conditions that were elaborated were different tensions in the drill string; the weight of the string and the bottom hole assembly could differ between the two sections, giving different depth measurements. Another possible explanation was that the radioactive markers that were placed inside the casing and offered some of the measurement points could have been displaced during the operations. Yet another scenario was that the rig was ballasted differently and therefore located higher or deeper in the water during the two different measurements. A fourth

----

5   12 ¼" (inches) and 8 ½" are standard diametrical dimensions for drill pipes. In a drilling operation, one starts out with a large diameter and ends up with a smaller. In a typical well, the series of sections will be 36", 26", 17 ½", 12¼" and 8½".

explanation was that different levels of the tide contributed to different distances to the measuring points.

None of these four scenarios could account for the whole depth difference of 4.5 metres, but a combination of them could possibly add up to the variance. A comment from one of the participants in the meeting was that this case showed how inaccurate the operations were, allowing all these sources of error to occur without anybody being able to account for them: 'if the effect of drill string tension and tide was automatically corrected for in the depth measurements, we could have started this debate on a much higher level.' This statement is pregnant with the ambivalence towards articulation work. On one side, it is appreciated, and met with expectations, expressed by a longing for 'debate on a much higher level'. On the other hand, the significance of the articulation work is partly neglected. At least, that is one interpretation of how the above statement and many of the other comments during and after the meeting commented that 'the effect of drill string tension and tide (should be) automatically corrected for in the depth measurements', while the value of the many scenarios and understandings that the discussions produced were not mentioned at all.

If the innovative strategy is focused mainly on decision support and automation and less on the inherent value of the articulation work it is designed to ease, some disregarded qualities of existing practices might get lost (Roth et al., 2006). If, for example, the review and execution of today's detailed operations procedures are replaced by an undebated, automated procedure, the variety of dependencies will be black-boxed and less accessible for inspection. Contingencies might thus materialise as huge surprises, instead of comprehensible coincidences. On the other hand, if the introduction of new work processes and technologies also aims at *strengthening* this type of articulation work, the effect of the innovations might be even better than anticipated, because the articulation work can be improved by fostering even more sophisticated understandings of the complex drilling operations.

8.5 Conclusion

Successful drilling is a result of collective work. Well-structured formal job descriptions, drilling programmes and job reports do not account for the situated articulation work involved in the actual accomplishment of work.

In this study, complexity is not seen as a static characteristic of a sociotechnical system; rather it has been treated more as a result of action than a determinant of action. The safety of the drilling organisation hinges, therefore, not so much on technical preconditions as would be assumed if the study was to be based on the presumptions of the NAT. At the same time, the mindset and social organisation that dominates the HRT would not be sufficient to account for the work processes within the rig team and its many associated human and non-human actors. Regarding complexity, the focus is on how it unfolds and what this process means for organisational safety. Sensitivity to operations and reluctance to simplify interpretations, which are important dimensions of HRO, is relevant although with a slightly different and more relative meaning.

A characteristic of the sociotechnical system of a drilling operation is its transient complexity. It is low during smooth drilling operations, when many actors and work processes are invisible to outsiders. When the opaque operations are made transparent, when dependencies are renegotiated and problems or unexpected events occur and are solved, complexity increases. A characteristic of the work of drilling organisations is its transient visibility. Many actors and much work are black-boxed and invisible most of the time, while they are made more evident in occasions when the operations are subject to closer inspection. Articulation work is a required labour input for the negotiation and resolving of complexity and visibility. But in addition, articulation work contributes to organisational safety through offering both a specific and a more generic understanding of the nature of drilling operations to the involved parties.

Emerging complexity represents both a potential crisis and an occasion for making work visible. One rationale of IO is to close the loop of decision-making, so that available information can be taken into consideration effectively and the action to be taken, preferably by automation, can be identified. In such an organisational change, there is a risk of handling complexity by bypassing work processes that are not subject to much attention, but that nevertheless contribute, potentially largely, to successful work in the organisation. What is proposed, therefore, is that articulation work should be seen not only as a time-consuming necessity which should be reduced as much as possible, but also as an activity with yields beyond the solving of ongoing problems. Articulation work brings to attention many aspects of drilling that are invisible most of the time. It produces scenarios of failure and expands imagination. Articulation work also offers scenarios where skills can be developed and practised.

To increase the potential of new technologies and work processes for not only better and faster, but also safer operations, the knowledge work of existing practices in the drilling organisation should be acknowledged and carefully accounted for. This study is a first step in that direction.

Chapter 9

Chasing Shared Understanding: Revisiting the Case of Divergent Depth Measurements[1]

Modern industries are characterised by a high degree of division of labour. The cooperation between different disciplines and expertises within organisations, and the coordination of the contributions of the different actors into collective achievements to ensure safe and efficient operations in risk-exposed industries is thus an important field for research. The topic has been treated in a range of workplace studies within many different domains such as aviation (Endsley, 1999; Hutchins, 1995a, b; Suchman, 1996a; Weick and Roberts, 1993), health care (Munkvold and Ellingsen, 2007; Tjora, 2000), London Underground control centres (Heath and Luff, 1992), the petroleum industry (Almklov, 2006; Hepsø, 2006; Rolland et al., 2006), and in more general workplace studies (Engeström and Middleton, 1996; Heath et al., 2000).

The recurrence of the theme of shared understanding in such studies indicates its relevance for collaborative work and Integrated Operations. However, the nature of shared understanding as such is not clear. A range of different conceptualisations of the phenomenon exist, and the role of these concepts in collaborative work is generally taken for granted and not scrutinised.

Lack of shared understanding is often pointed out as a main cause of failure (see e.g. Klein, 2005; MacMillan et al., 2004; Roth et al., 2006). Accidents are often shown to *coincide* with, and thus believed to be *caused by*, a breakdown in shared understanding. Conversely, a high degree of shared understanding is seen as a contributor to safe and efficient operations. The US Army thus states that 'shared situational awareness, coupled with the ability to conduct continuous operations, will allow Force XXI armies to observe, decide, and act faster, more correctly, and more precisely than their enemies' (TRADOC 1995: paragraph 1–2) and in connection with Integrated Operations, it is stated that 'shared understanding has a significant impact on the ability of teams to coordinate their work and perform well' (Grøtan et al., 2009: 2221). This straightforward coupling between shared understanding and safe and efficient collaborative work is problematic, because shared understanding is an underspecified phenomenon, and shared situation awareness, for example, is 'elusive and ill-defined, and does not lend itself easily

---

1   This chapter contains an analysis of the case presented in Section 2.2. The reader is advised to reread this case before continuing to read this analysis.

to traditional scientific evaluation' (Nofi, 2000: 71). Actually, an alignment of understanding and awareness is not always desirable:

> Agents within a system each hold their own situation awareness, which may be very different from [although compatible with] that of other agents ... We should not always hope for, or indeed want, sharing of this awareness, as different system agents have different purposes. (Stanton et al., 2006: 1288)

The writings on group think (Janis, 1972), conceptual slack (Schulman, 1993), requisite variety (Weick, 2007) and ambiguity (Antonsen, 2009b) are also a reminder that shared understanding is not necessarily a precondition for safe and efficient collaborative work.

The objective of this chapter is to explore different conceptualisations of shared understanding by applying them to a real case of collaborative work where the actors' goals and success criteria are not unanimous. The relations between the concepts will be investigated, as will their ability to explain and affect the safety and efficiency of collaborative work process. The results are believed by a large community of practitioners and researchers to be important to the further development of Integrated Operations, an operating regime whose safe and efficient collaboration is often held to be connected to shared understanding (see e.g. Grøtan et al., 2009; Kaarstad et al., 2009).

Different concepts of shared understanding – common ground, shared situation awareness and common information spaces (CIS) – are compared and applied to a case to explore their contribution to the cooperative work. The field of study is an onshore rig team within an international petroleum company. Being responsible for the offshore drilling operations, the rig team writes the drilling programme and follows up its execution. In practice, this work involves cooperation between the rig team and a range of other actors such as the offshore rig crew and the drilling contractor, onshore geology and reservoir experts and service companies with different types of specialist expertise that are needed in the different stages of a well project.

The case that will be subject to investigation has already been presented in Section 2.2. Before we proceed with the analysis of the case, we will briefly review the concept of shared understanding as it appears in the research literature.

9.1 Shared Understanding in the Literature

A characteristic feature of modern industries is the division of labour that renders possible a high degree of specialisation and the accomplishment of highly complex work. A challenging implication of this division of labour is the work related to coordinating the different contributions – to put together what has been divided. Two different approaches to describe and explain this type of work may be identified. One describes the cognitive processes involved in the work, emphasising these processes' distribution over humans and artefacts rather than being merely individual mental processes. Just as the cognitive processes of a sociotechnical system can be described as *distributed* (Artman and Garbis, 1998; Hutchins, 1995a, b), the way the system may have a common situational understanding is best described as a *distributed understanding*[2] (in contrast to overlapping). This approach will be revisited towards the end of the chapter, when the conclusions are drawn.

The other tradition, which is thoroughly explored here, focuses on *shared understanding* as a central entity for successful collaborative work. The terminology, however, is not uniform across the disciplines following this tradition; each discipline adopts its own concept and fills it with its own meaning. With slightly different connotations, the concepts are many; common understanding, team shared awareness, shared understanding, group situational awareness, shared cognition, team awareness, coherent tactical picture, common ground, shared workspace awareness, team cognition, shared mental models and common information spaces (see e.g. Bannon and Bødker, 1997; Nofi, 2000; Roth et al., 2006).

In the following, three distinct concepts for shared understanding will be reviewed: common ground, shared situation awareness and CIS. The selection is made with the purpose of covering different disciplines and different epistemological perspectives; a social constructive view, a realistic view and a sociotechnical, relativistic view. The selection thus covers some central aspects of knowledge and collaboration; the involved actors *perceive*, and they *share* their perceptions through *social* and *technical* interaction. Thereafter, an occurrence from the observation study that has already been reviewed in Section 2.2 will serve as a case for exploring how the different types of shared understanding relate to the coordination and accomplishment of the work.

It is important to note that it is not the qualities and pertinence of the different concepts as such that are investigated, but the way the concepts are used to explain the outcome of collaborative work.

9.1.1 Common Ground

The notion of common ground is rooted in Herbert Clark's contribution theory. It was coined to describe the way people achieve joint understanding, in the form of 'mutual knowledge, mutual beliefs, and mutual assumptions' (Clark and Brennan, 1991: 127) in the course of conversation. The process by which common ground is achieved between two or more participants is called a grounding process. The grounding criterion is met when 'the contributor and his or her partners mutually believe that the partners have understood what the contributor meant to a criterion sufficient for current purposes' (Clark and Brennan, 1991: 129).

---

2   Stanton et al (2006) calls it distributed situation awareness. Here distributed understanding is used to avoid unnecessary confusion with Endsley's (2000) concept.

The contribution theory that serves as the theoretical base for the formation of common ground is primarily a theory of discourse. That is, the focus is on the process of and the resources available for sharing already achieved understandings rather than the process of gaining new knowledge about the world. The significance of the media through which the communication between the participants takes place is evaluated only for media which are dedicated to communication support. This has been subject to the criticism that contribution theory and common ground does not take into account other aspects of communication and cooperation such as embodied phenomena and the material and social environment (Koschmann and LeBaron, 2003).

Common ground as a basis for cooperation involves a social constructivist approach to cooperative work. Basing cooperation on common ground implies much focus on *communication about the world* and little focus on *accounting for what makes up this world* (see e.g. Beers et al., 2005). It also builds on the assumption that the participants are informing each other, rather than creating new knowledge together; hence the social constructivist label to distinguish it from the more sociotechnical epistemology of common information spaces (see Section 9.1.3).

Clark and Brennan adapt Grice's (1975) principles of least effort in the grounding process into an adjusted principle of least cooperative effort; 'in conversation, the participants try to minimise their collaborative effort – the work that both do from the initiation of each contribution to its mutual acceptance' (Clark and Brennan, 1991: 135). This is obviously relevant for the cost-efficient achievement of common ground. How this common ground in turn affects the production of safe and efficient collaboration, however, is not well documented.

9.1.2 Shared Situation Awareness

The notion of shared situation awareness has been widely used in the field of human factors. Central contributions have come from the field of psychology, and aviation has been one of the industries where it has found application. It builds on the more generic concept of situation awareness, defined as 'the perception of elements in the environment within a volume of time and space, the comprehension of their meaning, and the projection of their status in the near future' (Endsley, 1988: 97). As an extension of this concept, shared situation awareness refers to the intersection of situation awareness among several actors. Endsley and Jones define shared situation awareness as 'the degree to which team members possess the same SA[3] on shared SA requirements' (Endsley and Jones, 1997: 37).

Situation awareness is based on a realistic epistemology, with a focus on objective perception of cues, comprehension of their meanings and projection to forecast future events presupposes a view where phenomena are out there, ready to be perceived, and that the challenge is to perceive them as precisely as possible.

---

3   Situation awareness.

Although widely used and accepted, the use of deficient situation awareness as a condition explaining why accidents happen (e.g. Aeronautica Civil de Colombia, 1996; National Transportation Safety Board, 1994) has been criticised by Billings (1996) and Dekker and Hollnagel (2004) for being deficient and tautological, and that the usage takes place 'without further specification of the psychological mechanism that might possibly be responsible for the observed behaviour – much less of how such mechanism could force the sequence of events toward its eventual outcome' (Dekker and Hollnagel, 2004: 79).

9.1.3 Common Information Spaces

In the field of computer-supported cooperative work, a concept that has received much attention is common information spaces (CIS). Although the concept is vaguely defined and has been used in many different ways, a definition that was coined in the early 1990s is still referred to by many contemporary writers (e.g. Bossen, 2002; Fields, 2005; Munkvold and Ellingsen, 2007; Rolland et al., 2006). A CIS comprises 'the artifacts that are accessible to a cooperative ensemble as well as the meaning attributed to these artifacts by the actors' (Schmidt and Bannon,1992: 21). A CIS is a space in which people can work cooperatively by

> maintaining a central archive of organizational information with some level of 'shared' agreement as to the meaning of this information (locally constructed), despite the marked differences concerning the origins and context of these information items. The space is constituted and maintained by different actors employing different conceptualizations and multiple decision making strategies, supported by technology. (Schmidt and Bannon, 1992: 16)

These spaces can apply in situations where people are co-present in time and space, or they can apply to settings where people work 'across time and space boundaries' (Bannon and Bødker, 1997: 2). In the latter case, the issue of stability is important because the CIS is to support distributed work in the course of time. People who work across large distances, perhaps not even aware of each other's presence and work contributions, need to relate to the same information and its attributed meanings. To support settings that differ with respect to boundaries of space and time, Bannon and Bødker (1997) suggest that the nature of the CIS needs to be dialectic. Within local communities of practice (Brown, 1991) the CIS might be open and malleable, allowing for interpretation and negotiation. CIS that serve cooperative work distributed over time and space, on the other hand, must allow for closure and immutability to function as immutable mobiles (Latour, 1987) that can be transported between locations.

The concept of CIS is hence more relativistic than those of common ground and shared situation awareness. Whereas situation awareness is referred to as ecological realism (Endsley, 2000; Flach, 1995), CIS resemble the sociotechnical relativistic epistemology as described by Latour (e.g. 1999b). The focus is on alignment of artefacts and the meanings ascribed to them. In this way, not only the understanding, but also the empirical world to understand, is based on construction rather than perception.

CIS does not avoid criticism. Among the characterisations is the view that 'the very notion of CIS is radically underspecified' (Randall, 2000: 17), and that 'there are a number of serious problems with the concept, and the way it is often used today. It would appear that rather than clarify matters, the label may only obfuscate' (Bannon, 2000: 1).

9.1.4 Shared Understanding and Integrated Operations

In the traditional petroleum industry, the division of labour is accompanied by a division of knowledge. A high degree of specialisation poses a challenge to the actors when decisions need to be based on a holistic understanding, and when the goals guiding the decisions differ between different disciplines. As a consequence, decisions that are taken within one discipline may be suboptimal or even harmful from the viewpoint of another discipline. This may typically be the case when two disciplines adopt different time horizons for their work and its outcome. One example is the drilling department whose time horizon includes only the construction of a well, and the subsurface disciplines whose time horizon is considerably wider, including not only the well construction but the whole lifetime of the well. Another issue is that work in the petroleum industry has traditionally been performed in a mainly offline modus, with a considerabletime delay between the creation of information and the use of this information for operative decisions.

These are among the challenges that are addressed by change processes that can currently be seen throughout the petroleum industry. The industry is undergoing a sociotechnical change process in the pursuit of more integrated operations. Technological and organisational efforts are being made to modernise an industry which is characterised by disintegration both in terms of knowledge and geographical allocations (OLF, 2005).

Closer integration of people, technology and information is one of the petroleum industry's responses to increasingly demanding operations with respect to reservoir characteristics, HSE[4] and profitability. Integrated Operations apply new technologies and new work processes to make data and information available to those who can make use of it, in real-time, to make collective and holistic decisions and to automate work processes. The technological innovations encompass a range of different aspects. New, more and better sensors will provide more information of higher quality. Enhanced signal transportation along wired drill-pipes will make a radically increased sampling rate and bit rate of subsurface data available in real-time. Enhanced information processing renders real-time updating of better models possible. New systems for monitoring operations will detect deviations that occur, or give warnings before they occur. Automation will,

---

4   Health, safety and environment.

to some extent, substitute human reasoning and action and standardise operations so that they are performed consistently and within safe operations windows (Iversen et al., 2006, 2007; Rommetveit et al., 2004, 2008b).

The new work processes involve closer integration of onshore and offshore personnel, and of different companies onshore. Better information and communication infrastructure will increase the use of video conferences and sharing of documents, pictures, graphs etc. The establishment of centralised expert centres will ensure that a limited number of experts can serve a large number of operations. A more global approach to operations and services will lead to the transcendence of time zones through a follow-the-sun allocation of tasks, presumptively resulting in faster, better and safer operations (Løwén et al., 2009; OLF, 2005; Ringstad and Andersen, 2006).

In light of the above, it is not surprising that shared understanding is a topic of much interest in relation to Integrated Operations (e.g. Andresen et al., 2006; Grøtan et al., 2009; Hepsø, 2006; Rolland et al., 2006; Skarholt et al., 2008; Tinmannsvik, 2008). A potential of technologies and work processes closely linked with Integrated Operations to promote shared understanding should be seen in relation to careful considerations of what shared understanding really means in practice, and what role it plays in the cooperative work. These issues are addressed here by following the advice of Hollnagel (2009) and Dekker (2006) to study normal work rather than failures. Hollnagel claims that 'if the probability of failure is as high as 10^{-4}, there are still 9,999 successes for every failure, hence a much better basis for learning' (Hollnagel, 2009: 83), therefore 'we should try to understand and explain the normal, rather than the exceptions' (Hollnagel, 2009: 97).

9.2 Method

The study of shared understanding in cooperative work was undertaken as an observation study in an international petroleum company. Over a period of eight weeks I joined an onshore rig team as an observer. The team is responsible for all drilling operations at one specific field. During the fieldwork, one well project – the drilling of one well – was observed from beginning to end.

The study offered an insight into the many different work processes within the team responsible for the offshore drilling operations. Access to the team was granted by the team leader (drilling superintendent), and allowed me to join the team in their everyday work in their open-plan office and their daily video conferences with the offshore rig crew and other collaborators. I was also granted nearly unrestricted admission to any ad hoc meetings that were held in the course of the operations.

During the observation study, which also involved participation in less formal settings such as coffee-break discussions and the daily lunch break, I became quite familiar with the members of the rig crew. As a supplement to the observation in professional work settings, this was a valuable trust-building socialisation that prepared the ground for the interviews that were conducted in the last part of the study, and for the gradual transition from pure observation to more participatory inquiries.

Four interviews were conducted, tape recorded and transcribed. The interviews were conducted in an informal, conversational manner that allowed the informants to focus on the topics that they themselves considered important in relation to cooperative work and shared understanding. Two of the interviewees were drilling engineers, a third was the team's HSE engineer and the fourth was a reservoir engineer. It should be noted that the reservoir engineer is not a core member of the rig team, but plays a central role in parts of the drilling operations. In addition to the four interviews related to the specific case, interviews and observations from other parts of the study have also been a useful resource for developing a richer understanding both of the case and of the role of shared understanding.

Apart from conducting observation and interviews, the author was also given access to the company's database where information about the organisation and its work processes, best practices and drilling programmes could be accessed. Considerable time was therefore also spent on literature review which helped contextualise the technical information and make it intelligible. With respect to making information from the professional petroleum domain intelligible to a social scientist, my professional background as an offshore mud-logging geologist was of course also extremely advantageous for gaining a relevant understanding of the rig team's work.

When not working individually in the open-plan office, the rig team works closely together with the offshore rig crew and offshore and onshore representatives from service companies that deliver equipment and services for the operations. It was thus not only the rig team that was studied, but a distributed, loosely coupled organisation that was constituted differently from meeting to meeting, depending on the meeting agendas.

As often as possible, morning meetings were observed. Frequent attendance at these meetings offered the author a regular update of the drilling status. The same was the case for meetings where detailed operating procedures were reviewed before they were carried out. These meetings were held prior to special operations such as running of casing, cementing jobs, sidekicks etc. In addition, ad hoc meetings that were initiated by contingent events were especially interesting, since these meetings represented occasions where drilling problems were elaborated on the spot, with minimal preceding alignment of viewpoints between people and across disciplines.

In the regular morning meetings between the onshore rig team and the offshore rig crew, there was typically only a basic, minimum attendance. Offshore, this included the company man, the company drilling engineer, the toolpusher and the offshore installation manager.[5] Depending on the operations, a geologist and a

---

5   The company man is the representative of the oil company, while toolpusher designates the supervisor for the drilling contractor.

reservoir engineer were also present. Onshore, the standard staff consisted of the drilling superintendent, an HSE engineer, a logistics engineer, a drilling engineer and the leading drilling engineer. Additionally, operation-specific personnel including a geologist, a petrophysicist, a reservoir engineer and a drilling coordinator from a service company were also present. Furthermore, the onshore supply base was always present as a third party in the video conference. These different types of meetings are important arenas for the negotiation and sharing of information and knowledge, and they thus offered valuable empirical data to the case.

The case study draws on different traditions within qualitative methods. As the research question and the themes in focus were not formulated clearly in advance of the observation study, but rather were highlighted by the observations and the interviews, the approach bears resemblance to Glaser and Strauss' (1967) grounded theory. Exactly what was looked for in the data material was highly influenced by the material itself. Garfinkel's (1967) ethnomethodology is another source of inspiration, directing the main focus on the informants' own descriptions and perceptions of their work rather than the scientist's interpretation of it.

As a central part of the method, informants have been involved as discussion partners in the rewriting process. Getting the details of the case right was crucial for elaborating the case in a relevant manner.

9.3 The Meaning of Uncertainty to Different Disciplines[6]

The drilling engineers represent the organisational owners of depth measurements and positioning of the well. A reservoir engineer explained that:

> These measurements are vitally important to the drilling engineers to determine the exact well path only after a well is constructed. Controlling this means that future wells can be planned for with sufficient margins in order to avoid collisions with older wells. To the reservoir engineers, on the other hand, the significance of real-time positions is vital. The accuracy and reliability of measurements are decisive, and they are so during drilling.

The informant continues to explain that the view among the reservoir engineers is that the drilling engineers have much lower requirements for accuracy in the drilling operations than the reservoir engineers have, and that the drilling contractors are known to be approximate about depth measurements. It is also a general view among drilling engineers that reservoir engineers often demand accuracies that are meaningless given the reliabilities of the methods of measurement. The informant used the case as an example; the reservoir engineers want to investigate the reason for the measurement deviances in order to reduce the deviances to decimetres,

---

6   Cf. the case in Section 2.2.

while the uncertainty of the measurements exceeds those limits by far. A drilling engineer emphasises that 'some degree of data uncertainty is impossible to avoid. We just have to live with that.' The main difference in the way drilling engineers and reservoir engineers relate to the uncertainties could perhaps be illustrated by the following quote from a reservoir engineer: 'when you talk to drilling people, there are very few who want to hear about uncertainties. Drilling people only want a yes or a no.' Understanding the different significance of uncertainties to different disciplines is important for understanding what the measurement discrepancies in the case mean to the different actors.

9.4 Discussion[7]

In the introduction of this chapter it was claimed that poor performance and accidents are often held to be caused by a lack of shared understanding, without accounting for the role shared understanding actually plays in normal, collaborative work. By simply showing the correlation between accidents and breakdown in shared understanding, and not accounting for the *causal* connections between shared understanding and successful work, accident investigators run the risk of drawing incomplete or even wrong conclusions with respect to what *caused* the accident. In the case, none of the three conceptualisations of shared understanding are identified as suitable models for the driving forces of successful collaboration. To the contrary, shared understanding does *not* enter into the description of the work. This is indeed surprising if lack of shared understanding is believed to unambiguously affect the collaborative work and its result in a negative way.

Based on the case, the subsequent sections elaborate in which way shared understanding is relevant to the work of the drilling engineers and the reservoir engineers. In order to understand the effect of shared understanding, such an analysis must be undertaken in accordance with specific and unambiguous definitions. In addition, the limits of the concepts' explanatory power must be acknowledged. Since each concept covers a non-exhaustive part of shared understanding, a central point must be to identify what is outside of each concept's scope, and what the implications are when different concepts are incompatible within the same area of application.

9.4.1 Handling the Uncertainties

In the course of the case, the participants elaborated on their different requirements with respect to issue A.[8] The reservoir engineers argued for an investigation of the measurements discrepancy in order to establish an undisputed understanding of the true state of the well. Such a realist decision-making strategy might well end

---

7   Rereading the case that was presented in Section 2.2 at this point will be useful.
8   The depth discrepancy.

up with a shared situation awareness. Although the epistemological preconditions for situation awareness are not clearly defined, the statement that 'it is entirely possible to have perfect SA, yet make an incorrect decision' (Endsley, 2000: 8) points towards a realist epistemology. It is hard to think of a perfect situation awareness and a wrong decision without relying on the correspondence theory of truth (Latour, 1999b). However, correspondence is not an unambiguous requirement in drilling operations. The goals of a discipline influence the state of affairs. The reservoir engineers' wish to calculate exactly the stretch effect of the drill string on the measured depth illustrates this; from the drilling engineers' point of view this was not relevant since this stretch of the drill string under any circumstance would be much smaller than the overall depth uncertainty. What was considered as shared situation awareness by the reservoir engineers was considered as an illusion by the drilling engineers.

The strategy suggested by the drilling engineers was to abandon the ambition of acquiring any absolute points of reference and just keep future measurements relative to the last measurement, undertaking the forthcoming measurement in the exact same way with identical equipment as in the last measurement. In that way many uncertainties would become irrelevant; if they had missed out an element of the drill string, they would 'miss it out' again by applying the same drill string configuration. If the radioactive markers were out of position, they would still be at the same position in the next run, since the casing would not have moved in the meantime. The identification of top Garn formation would follow the same interpretation principles. Hence, by using the measurements of the last section as a reference point for the next, it would be of less interest whether these measurements were *correct* or not, given that the tide and the ballasting during that last run was accounted for. This approach would involve an expansion of a CIS into the near future, leaving the configuration of tools and the interpretation of information unaltered from one situation to the next. This strategy suggests a more relativistic form of understanding than the first approach. Whereas the first approach is based on a form of understanding based on 'objective facts', the second approach is based on a pragmatic and a constructivist understanding, on the construction of a CIS.

The choice of approach has consequences beyond the actual well project. The first approach will contribute to an understanding not only of the actual reservoir but also to the wider geomodel of the whole field. The second approach might serve the accuracy of the singular well better than the first approach, and is not based on the accuracy of measurements as much as on the reliability of *relationships* between measurements. The drawback is that the value potential will not be transferred to the wider geomodel, as in the first approach, because the location-specific context of the uncertainty cannot be transferred to operations situated elsewhere.

Issue A (the depth discrepancy) was not discussed in isolation from other issues. One such issue was issue B,[9] since 'the pressure gradient could be used to determine the oil/water contact and thus make oil/water contact an external depth reference that can be reused in later operations' (reservoir engineer). In addition, 'the gradient could give important information about the reservoir with respect to faults and prospects for future production' (same informant). Issue B could thus be understood as an integrated part of issue A in light of a realist strategy, and it could contribute to the approach preferred by the reservoir engineers. On the other hand, as already noted, the drilling engineers were not enthusiastic about pursuing issue B because it would involve extra costs that would not contribute to *their* understanding of the state of affairs.

With the different perspectives on how to handle the measurement discrepancies with respect to the local/global and the short- and long-term perspective, perhaps common ground could be the missing link needed to negotiate the optimal solution? As we shall see in the following sections, neither common ground, shared situation awareness nor CIS, as concepts for shared understanding, are fully adequate to support the work and the decisions needed to handle the discrepancies in the case.

9.4.2 Not Just an Issue of Communication (Analysing Common Ground)

Achieving common ground, or 'mutual knowledge, mutual beliefs, and mutual assumptions' (Clark and Brennan, 1991: 127) with respect to the ongoing operations does not seem to be the primary challenge for the cooperative work. The drilling engineers and the reservoir engineers are well aware of each other's perspectives and needs with respect to data accuracy. The elaboration by a drilling engineer on the limitations with regard to reliability of the measurements wanted by the reservoir engineers shows that the coordination of decisions related to the case depends more on trade-offs between accuracy and reliability than on obtaining common ground. Mutual knowledge, beliefs and assumptions do not concern the specific situation as much as they concern the larger goal, which is presented in the drilling programme of the actual well as 'to drain as much oil ... as possible from the ... reservoir'. Whether the decision is based on the first approach of investigating the difference in measurements to establish an undisputed understanding of the true state of the well or the second approach of abandoning the search for absolute points of reference and rather keep future measurements relative to measurements performed with identical equipment, hinges more on a clear formulation of an overarching goal of the operations than on mutual knowledge, beliefs and assumptions – common ground – with respect to the case of the depth measurements discrepancies. It is on the overarching level that the decisions can rely on a social construction. In specific situations such as the present case, the depth references must also be based on physical measurements of depth and on a specific configuration of the drill string.

---

9   The number of pressure points.

9.4.3 Objectivity is Relative (Analysing Shared Situation Awareness)

The perception of issue A and issue B, and their interdependence, depends on whether it is based on a short- or a long-term perspective. According to Endsley and Jones (1997: 20), 'situation awareness is highly impacted by a crew member's goal and expectations'. However, since there are at least two different solutions to the case problem, one unique reference against which to evaluate its correctness does not exist. The solution cannot simply be perceived, it has to be constructed; the participants must choose how the depth measurements should be handled. The first, objectivistic approach, which could be justified with reference to situation awareness, is the one favoured by the reservoir engineers. The advantage with this approach is that it is relevant for the larger field and hence can add to the long-term value creation. The paradox is that due to the limited accuracy of measurement (see Section 9.3) such an approach can turn out to be non-optimal in the short-term perspective, when only the actual well is accounted for. Thus, a non-contextual evaluation of this specific well project and the team that undertook it could be unfavourable. Furthermore, because the actual drilling process is managed by the drilling engineers, not by the reservoir engineers, it is difficult to argue for such an approach. As stated by a drilling engineer: 'Some degree of data uncertainty is impossible to avoid. We just have to live with that.'

9.4.4 Relativity has a Limited Reach (Analysing CIS)

More than a collective mental state corresponding to an objective reality, the challenge of the rig team is to negotiate and to *construct* the well with the information and tools that are available. Seeing the activity to regain control over the drilling operations as an effort to construct a CIS engages a variety of extra resources, such as the drill string configuration and relative measurements available to the participants. With such a perspective, relativity and context is introduced as an ingredient of knowledge. Besides, the epistemology that makes such a perspective possible allows for *construction* to supplement the less powerful process of *perception*. This sociotechnical construction is more comprehensive than the social construction discussed previously, but not necessarily fully adequate. Instead of viewing the measurement discrepancies as a problem of communication (common ground) or shared perception (shared situation awareness), this approach would find the challenge to be an issue of actively combining the available social and technical resources (CIS) into a stable configuration. By including the *context* of the measurements of different sections of the well in the measurements, the measurements become *relative*. In this case, the context is made up of the identified six potential sources of error highlighted in Section 2.2. To include the context means that the measurements of depths in one section are transformed from an isolated, objective measurement into a relative measurement that includes a drill string configuration both of the current section and the sections to come. The implication of such an approach is, however, not unproblematic, as shown by this case; whereas it may increase the reliability of the depth measurements of the different sections in the *current* well, it may reduce the reliability of the same measurements in *future* operations where the stability of the CIS has not been maintained because the context is impossible to reproduce.

Here, the tension between local and global is made current. This tension has been emphasised by others (e.g. Rolland et al., 2006). The dialectic character of a CIS suggests that it is a temporal entity whose unambiguous configuration is problematic to transport to another location in a future operation. In this case, an extension of the CIS would depend on reproducing the potential sources of uncertainty at another location. However, since every location and every operation need equipment and drilling strategies that are adapted to the local conditions, such an extension of the CIS would be based on an illusion.

9.5 From Shared Understanding to Distributed Cognition. Conclusions

A case of measurement discrepancies has been reviewed in light of different concepts of shared understanding. The review has been undertaken to investigate how shared understanding can play an active part in negotiating a solution to the problem. What the case shows, surprisingly, is that shared understanding is *not* a requisite in this work. The different concepts of shared understanding investigated *may* contribute to solutions that are valid for a specific time/space combination,[10] but when it comes to supporting solutions that are valid across a wider time/space domain,[11] they are shown to be less functional, even contradictory. The limited functionality can be traced back to each concept's limited area of application (cf. Sections 9.4.2–9.4.4). The contradictory aspect has to do with the incompatible epistemologies that the different concepts of shared understanding are based on (cf. Sections 9.1.1–9.1.3 and 9.4.2–9.4.4). These limitations could be compared to the limitations of a map projection where a three-dimensional geographic relationship is transformed to a two-dimensional surface; a map projection may be conform (areas are correctly represented) *or* equivalent (areas are correctly represented) *or* equidistant (distances are correctly represented), but it cannot satisfy the demands of more than one projection at a time.

Shared understanding is a central phenomenon in collaborative work, and this case study does not challenge its position as a descriptive phenomenon as such. What it challenges is the position of shared understanding as a causal agent in collaborative work where the goals of the different actors are not unanimous. In such work, where there is a need for negotiations and mutual adaptation of the empirical world to the different goals and available methodologies, the shortcomings of shared understanding must be acknowledged. Rather than *shared*, the relevant collective understanding could be described as *distributed*, and

---

10   The specific well.
11   Future wells and the larger geomodel.

although this distributed understanding in itself may have limited causal powers, it is closely connected to the more powerful concept of distributed cognition whose significance is well documented by other authors (e.g. Artman and Garbis, 1998; Hutchins, 1995a, b) and which this case also may serve as an illustration of.

9.6 Epilogue

The well construction process was brought to an end with a satisfactory result in spite of the uncertainties with respect to the depth measurements. Surely, the final well report summing up the operation more than a year later eventually offers some kind of shared understanding of the well project. However, the insignificance of this shared understanding for solving the case problem is characterised by its post hoc character and the fact that the final well report does mention the case at all.

Chapter 10

Challenging Controversies – A Prospective Analysis of the Implications of Integrated Operations

Before we draw any categorical conclusions on the basis of the cases, let us see how the stakeholders of Integrated Operations themselves relate to the issue when they reflect broadly and freely on it, without relating it to any case that may challenge the generic value of their reflections. In this chapter, the analysis is carried out by the stakeholders themselves; the author's role has been reduced to that of a moderator who systematises the reflections so that they constitute a holistic picture together with the more case-specific findings.

10.1 The Arrangement of the Study

This part of the study is based on a series of qualitative interviews. Thirty-four employees from three different oil and gas companies, two research institutes and three technology developers were interviewed on the topic of Integrated Operations and its perceived influence on the work processes of drilling operations. The informants were identified through a mix of snowball recruitment and advice from a few key researchers with much experience and an extensive network within the IO domain. The professions of the informants are listed in Table 10.1.

The qualitative interviews were conducted in an informal manner and revolved around the informants' own comprehension of their work practices. The informants were allowed to elaborate on the topics they found relevant to their own work and, if necessary, to drift along seemingly idiosyncratic lines that might cover fields that otherwise would not be thought of by the interviewer.

As a point of departure, the informants were asked to give an account of the challenges of their work. Thereafter, they were asked to elaborate on how they expected the three IO tools presented in Section 7.5.1 to influence both their own work and the cooperative work involved in the drilling operations: who will be affected, in what way, and with what results? The majority of the interviews were recorded and transcribed. Then the search for themes that could illuminate the relationship between Integrated Operations and safety began.

The views on the challenges of drilling operations were far from uniform; the different perspectives were actually a common feature of the themes. There is a potential for confusion when a search for clear tendencies is rewarded only

Table 10.1   Number of informants from the different professions

Informants' professions	Number
Drilling engineers	6
Drilling plan leaders	3
Drilling superintendents	3
Drilling data analysts	2
Geologists	3
Reservoir engineers	4
IO advisor/manager	2
IO technology developers	8
Researchers drilling and well	2
HSE engineer	1
Total	**34**

with ambiguities. However, by mapping the different perspectives, or, using the vocabulary of Latour (2005), the *controversies*, and by letting 'the actors deploy the full range of controversies in which they are immersed' (Latour, 2005: 23), ambiguities could be turned into assets. Instead of letting them confuse the study, the controversies were regarded as a central resource for understanding the complex nature of integrated drilling operations.

A number of relevant topics were identified. These topics were, in turn, compared and grouped, resulting in a smaller number of themes. The subsequent readings of the data material were guided by these themes. Through an alternation between empirical data and themes of concern, the themes of controversy gradually consolidated and thus became adequate elements for the prospective analysis.

But before I present the picture drawn mainly by these IO stakeholders, let me show how our toolbox of theories and methodologies was applied to assist in the picture composition.

10.2 Opening the Safety Science Toolbox

A core safety challenge in drilling operations is to control the drilling process by retrieving the necessary data, interpreting it and comparing it to existing plans and models of the operations. With more and better sensors and improved transmission channels in the drill string, the amount of data in drilling operations have increased significantly in recent years. However, the increased production of data is not synonymous with increased control.

Since several reports (e.g. OLF, 2003, 2005; Stortinget, 2004) during the first half of this decennium first addressed an incipient change process in the petroleum industry, the acknowledgement of a new operating regime now encompasses most of the actors in the industry. This new operating regime implies strategies to 'use ... information technology to change work processes to achieve better decisions, control equipment and processes from remote locations' (Stortinget, 2004: 34).

Improved safety is a central aim of Integrated Operations, and although it is a general assumption that better integration will lead to improved safety (Moltu and Nærheim, 2010; OLF, 2005), the conditions for this effect to occur are not thoroughly elaborated (but see e.g. Albrechtsen, 2010; Grøtan et al., 2010; Ringstad and Andersen, 2006, 2007). The relation between the collection and handling of information, the technologies and work processes applied, the cognitive processes and the effect on safety is complex and demanding to analyse. One way of obtaining more precise analyses is to undertake domain-specific empirical investigations. The aim of this study is to explore how three prospective tools for interpretation, diagnosis and automation of the drilling process specific aspects of Integrated Operations will affect the safety of the execution phase of the well construction process.

Among the characteristics of Integrated Operations are increased instrumentation, new and more associations of heterogeneous elements (humans and technologies, different professions, different locations) and cognitive processes that are more distributed in the sociotechnical environment than before. Actor-Network Theory (ANT) is a theoretical and methodological tool that is well suited for exploring the impacts of such characteristics. But let us first briefly recapitulate and keep our theoretical toolbox open for a while.

The highly influential approaches of Normal Accident Theory (NAT) (Perrow, 1984) and High Reliability Organisations (HRO) (Weick, 1987; Weick and Roberts, 1993) disagree on the possibility of controlling complex sociotechnical systems. One source of disagreement is the organisation's ability to alternate between a centralised and a decentralised control modus. Perrow points out that the control of organisations with tight couplings needs to be centralised while the control of interactively complex organisations needs to be decentralised. As a consequence, he claims, the demands for organisations that are both tightly coupled and interactively complex are incompatible. HRO researchers, on the other hand, argue that some organisations are able to reconfigure quickly and adapt to the control demands in crisis situations. Following a more generic organisational research tradition, Mintzberg's (1979) descriptions of different types of control, or coordination, challenges the dichotomous distinction between centralised and decentralised control. Mintzberg's approach is, however, structural rather than cognitive.

Latour's (2003a) interpretation of Beck's (1992) notion of risk offers an alternative entry to the study of risk and safety in sociotechnical systems by applying ANT:

[Risk] does not mean that we run more dangers than before, but that we are now *entangled*, whereas the modernist dream was to disentangle us from the morass of the past. A perfect translation of 'risk' is the word *network* in the ANT sense, referring to whatever deviates from the straight path of reason and of control to trace a labyrinth, a maze of unexpected associations between heterogeneous elements, each of which acts as a mediator and no longer as a mere compliant intermediary. (Latour, 2003a: 36)

In the ANT perspective, the contrast between humans and technology, and between centralisation and decentralisation, is replaced by the network that may be described by mapping the associations between heterogeneous elements and the way they mediate each other and the system.[1] Such an approach also accounts for the system's cognitive processes.

The descriptions of sociotechnical systems that are still in the making are especially challenging to produce because the traces of the associations have still not manifested. On the other hand, the advantage is that the controversies of the system have not yet been black boxed and a description of these controversies may thus reveal processes that tend to be invisible at a later stage (Latour, 1987, 2005). ANT has been an especially fruitful approach in science and technology studies (Latour, 1986, 1987, 2004a).

The field of distributed cognition (Hutchins, 1995a) is another research tradition that shares Latour's view on cognition;[2] cognition should not be understood as a purely internal mental process within the brains of individual human beings, but as the 'propagation of representations through various media, which are coordinated by a very lightly equipped human subject working in a group, inside a culture, with many artefacts and who might have internalized some parts of the process' (Hutchins, 1995a: 316).

The term 'distributed cognition' is specifically associated with Hutchins' writings. Using different terms, Latour also describes cognition as an external process largely influenced by material and social resources. When describing the visible traces of cognition he refers to 'a cascade of mobile inscriptions' (Latour, 1986: 28), to the creation of representations so mobile that they can be circulated and combined across a network of actors.

Hutchins (1995b) elaborates on the nature of cognitive work in the intense and detailed description of the activities inside a cockpit during the landing of a plane. Latour and Woolgar (1986) used the external perspective on cognition to account for work in The Salk Institute for Biological Studies. Although the large amount of paperwork makes the laboratory look like an administrative agency, Latour and Woolgar (1986) show that the paperwork is a natural part of both laboratory work and scientific, cognitive work in general. Latour also discusses the role of

---

1   Cf. Chapter 7.

2   For an elaboration of the similarities (and differences) between Latour and Hutchins, see Latour (1996d) and Giere and Moffatt (2003).

paperwork in connection with centres of calculation (Latour, 1986, 1987). Centres of calculation are locations or constellations where the necessary resources for performing calculations by combining different types of representations are gathered. There is reason to believe that IO tools may empower different actors in the drilling organisation to access and combine information to an extent that until now has been impossible; thus potentially giving them status as centres of calculation.

10.3 Controversies of Drilling Operations

The interviews did not provide any unambiguous answer to the question of how the tools will affect the safety of the drilling operations. What *were* revealed, however, were some themes, or controversies – some characteristic of the operations.

10.3.1 Information Overload Versus Information Availability and Quality

One aspect of drilling operations that many of the informants pointed to as characteristic was the amount, availability and quality of drilling data and background geological data. One informant described the situation with respect to availability of data:

> We have serious problems at the moment with the data or the information of the data. The information to make a good decision is usually there somewhere. But it's very hard to get it, sort of, in a timely and orderly way, before problematic situations develop into something serious.

The problem is not only that it is difficult to get hold of the information; information overload can be even more problematic than inaccessibility of data, as indicated by the same informant: 'You've got lots of systems, some of them talk to each other, some of them don't, so you can have several different depths, or you can have several different pressures; you can have calculated pressures, you can have measured pressures.'

Both inaccessibility and excess of information represent a threat to the uniqueness of reference that is often held as the ideal for controlled operations. Adding to the problem is the quality of much of the data, which was described as very low by several of the informants due to inappropriate measurement regimes. One informant puts it this way:

> I have mentioned the mud parameters; the way it is done today it is ridiculous. You bring a cup, a little cup of one or two decilitres, to the mud pit of 50 or 70 thousand litres. You take a sample and you do some tests and measurements. You do this four times during twenty-four hours, and then you use these measurements to change the mud parameters ... So, it is definitely about time

that we get automated and continuous measurements of the mud parameters. (Original quote in Norwegian)

Descriptions of inaccessibility of data, excess of data or data of dubious quality permeated the series of interviews. One of the informant sums up these challenges in what he believes to be the solution:

> To increase the accuracy and quality of the data, the accessibility needs to be increased. This is happening today, with wired drill pipe and increased instrumentation downhole. Thus we get more information from the well during drilling. With new software packages[3] we can also perform more real-time analyses instead of today's post-analyses based on memory-logs. (Original quote in Norwegian)

The expectation is that more sensors, better infrastructure and better interpretative and automation software will provide a continuously updated and accurate understanding of the drilling process and thus lead to increased safety. However, the tools' impact on the quantity, quality, accessibility and coherence of data is not unambiguous. Two aspects were mentioned by several informants: first, technology-supported production, transport and interpretation of data do not guarantee an increased accessibility of data to all stakeholders. Even today, the problem with data accessibility is not only a technological problem, it is also a political one. One informant working in an expert support centre of a Norwegian oil and gas company exemplified this by referring to an international operation, a joint venture led by another oil and gas company. The informant recounted:

> [Our company] had to pay a licence fee for every additional user that retrieved data outside of an initial list, so for four or five of us in the drilling room to get access to that data was gonna cost some amount of money per day, and we got access for maybe a week, and then the company said 'no'. So we weren't really able to get much data.

Second, software that integrates multiple data sources is vulnerable to the multiplication of uncertainties inherent in each data source:

> If you're drilling, and drilling has got certain uncertainties, say, with depth, the uncertainty is probably two per thousand, so that could be your uncertainty, plus two, minus two, in a thousand metres. But the geological uncertainty, say in a formation, could be plus/minus five metres, right, and then you have the modelling of the reservoir that has an even different uncertainty. And then you have the seismic uncertainty, sometimes that's very large ... You have to pull

---

3   Such software packages are the essence of Alpha-Drill, Beta-Drill and Gamma-Drill, the same tools that were described in Section 7.5.1.

this together. I think that's a big chance the more you integrate stuff, the more that can, sort of, manifest it into a more serious situation. I think that's a big challenge. (IO advisor)

10.3.2 Centralised Versus Decentralised Control

Another characteristic of drilling operations highlighted by many informants is the organisation of knowledge work within the organisation. A central point is the comprehensive outsourcing of analysing work to service companies. 'If we're going to do a wireline operation [service company A] is doing the simulations for us. And [service company B] are doing all simulation connected with drilling. Everything is outsourced.' (Original quote in Norwegian)

This division of labour, where different service companies independently perform the different drilling data analyses and where the rig-team's responsibility to coordinate the results from the analyses and deploy them in decisions based on the whole picture, is a source of concern mentioned by many of the informants, one of them quoted below:

> The different actors that deliver only a part of the chain ... they don't see the whole picture. They can recommend things that are not optimal for another actor, or for the overall operation. Take cementing, for example. [Service company B] may recommend that we cement as much as possible on the outside of the casing. They don't worry that we thus risk losing cement to the formation if the formation is not sufficiently strong. (Original quote in Norwegian)

The rig teams' task of coordinating the work of the specialists and utilising and integrating their data without having performed the calculations and interpretations themselves and without being familiar with the premises, assumptions and trade-offs involved, is challenging. Although the rig team has access to some simulation tools, there is a lack of both time and competence to apply them regularly, as stated by a rig team leader:

> We also have [the tools], but we don't have the time. Neither do we have the competence to use them satisfactorily. We have simulation tools for torque and drag, and for hydraulics. We also have others, but it is mainly those that we use. But we don't use them much. And it is only a few that use them. (Original quote in Norwegian)

Also, there is a need for simulation tools that integrate the different types of data that are necessary to solve more complex problems, as expressed by an informant: 'I think we are doing too little work on simulations. I think we should do more of that. Having special tools for such simulations could be very, very useful.'

The outsourcing of coordination work to other experts could be interpreted as a response to the problem of coordinating contributions with underlying assumptions and trade-offs that are not sufficiently known. Both the oil companies and service companies have such expert services that the rig teams can utilise. One informant elaborates:

> [Service company A] has expertise on drilling, drilling parameters, drilling optimisation, vibrations, tools, the drilling technology ... Or if we get problems with losses ... then our own company's expert centre can be contacted. They have a more general support, while [service company A] is more specialised on optimisation of the drilling process. (Original quote in Norwegian)

However, none of these centres are manned or equipped or given the mandate to administer the whole of the operations on a continuous basis. This overall responsibility, which resides with the rig team, involves more than the engineering skills available in such centres.

10.3.3 Engineering Versus Paperwork

The division of labour, as it appears on organisational charts, often seems to suggest that the nature of engineering work is fundamentally different from administrative work, and that work can thus be organised as a workflow of discrete tasks where the engineering work of calculation and the administrative paperwork are kept apart. In practice, this is far from the case. A newly employed engineer was thus surprised that paperwork should take up so much of his time:

> What struck me when I started as an engineer was that we outsource a lot, I mean, you don't perform the calculations ourselves any more, although you are an engineer ... So, it struck me that it is all about coordination. And not so much about engineering, in a way. (Original quote in Norwegian)

The time left for what he thought of as engineering work is very limited:

> It's a pretty hectic day. If you want to run a simulation, for example, during the operations, you need data. And those data take time to gather. And then you have to enter them into your system, you need to run the simulations with different uncertainties and with alternative input parameters. We don't have much time for such work. Paperwork takes time, you know.

Although more experienced engineers get used to, and accept, that paperwork is an integrated part of engineering work, the idea of engineering work as confined to simulations, calculations and interpretation of geological data and drilling data is still a dominant view. Here is how a rig-team leader evaluates the composition of his colleagues' work tasks:

Unfortunately, they use a lot of time on logistics. Get hold of equipment, for example. Such things are what take most of the time. Not engineering work.

What kind of tasks do you wish they could spend more time on? (Question from interviewer)

Engineering, monitoring the hydraulics, that is, the friction losses, monitoring the torque and drag, how the drill string behaves in the well compared to expected behaviour, those kinds of things. (Original quotes in Norwegian)

Many of the informants pointed to this mismatch between their own idea of engineering work and the experienced reality of administration and paperwork. IO tools for interpretation, diagnosis and automation were also mentioned as a means to reintroduce the *engineered control* over the operations. As one informant said:

This is what we are trying to change now. They want us to use a lot of simulation tools to go back to, you know, what we learned in school. All the simulations and interpretations *they* (the service companies) are doing, we shall do, too.

The engineering work of data gathering, simulation and interpretation is deeply intertwined with administrative work. This work was not part of the geoscientists curriculum at the university, nor is it what the new drilling engineers at the rig teams expect to spend much time on when they start their career. Still, administrative work is a core activity in drilling operations.

10.3.4 Standardised Versus Unique Interpretation of Data

The last controversy evident in the interviews was that of standardised versus unique interpretation of data. Some informants expressed a need for more instrumentation and standardisation of the drilling process, at the expense of the human experience and 'feelings'; the trustworthiness of which they devaluated. This is how one informant put it:

We want to make ourselves independent from feelings. We want to proceed to *engineering* operations. I often say that ideally we should make ourselves independent from experience. Experience is not an unambiguous entity. Different participants present in the same situations could gain different experiences. That's one thing. Another thing is that misinterpreted experience can be more harmful than useful. My opinion is therefore, that it is better to base the operations on real-time physical measurements and a trustworthy instrumented, model-based understanding of the state of affairs. (Original quote in Norwegian)

The same informant acknowledged that such a regime would require higher data quality than what is available in today's operations (cf. Section 10.3.1).

Vulnerability associated with low data quality would thus render necessary a critical review of all parts of the operations where data are generated, he continued: 'An automated system can only respond to input data, and it is therefore important that they are very accurate.'

Contrasting this, other informants argued that experience and sensitivity to the uniqueness of situations and data was something that was reserved to human beings and was thus threatened with being undermined by the introduction of machines into the cognitive domain. One informant argued by describing an experienced colleague on his rig team:

> Well, this guy has 25 years of experience ... It is hard to quantify, but I would say that the experience is very important in this work. To see the connections. Some of it can be standardised, but some of it is not possible to standardise in a software.

A third perspective on this theme was that human cognition is actually dependent on and empowered by tools that render possible controlling, monitoring and interpreting larger amounts of information than the engineers can handle today. One informant, working with the development of a case-based reasoning drilling software (Beta-Drill), argued:

> They [the drilling engineers] are getting real-time data from operations, that is, they are getting sensor data from the operation as it progresses, and they don't necessary know how to handle them. Also, when the real-time data turns into historical data, there are ways of utilising those data that have not been adopted. They acknowledge that they have a lot of unused data. So, we are trying to establish a connection between real-time data and historical data so that we can use historical data to recognise a situation from the real-time data, and use the experience from historical situations to solve present problems.

As cognitive capacity is empowered by tools that are capable of processing data quantities that human beings alone can only dream of, the border between standardisation and uniqueness gets blurred. A software-based management of information depends on predefined algorithms, and in that way facilitates standardisation. On the other hand, if the algorithms are sophisticated and the input data is of high quantity and quality, the nuances of singular situations could produce unique, tailor-made solutions. This, according to one informant, could render possible the safe and efficient drilling of wells that are more complex and demanding than are possible to manage with today's methods:

> I think it will lead to, in some ways, more standardisation. In *some* ways. Particularly on more commercial, standard wells. But it will also open the window or the opportunity to drill less standard, more complex wells. (Same informant as above)

10.4 Elaborating the Controversies – from Contradictions to Assets

The controversies identified are not only descriptive for today's drilling operations. In change processes, many features of the 'old' regime have a tendency to deposit rather than disappear (Hanseth and Monteiro, 1998). Thus, as the operations take on a more integrated character, the controversies are likely to be affected, but not resolved.

A common reference of the four controversies is the issue of development and administration of knowledge in sociotechnical work. As the following discussion will show, the answer to the question of what such knowledge development and administration depends upon, or how cognition works, also indicates the answer to how Integrated Operations might affect the safety of drilling operations. If cognition depends on its distribution over human and non-human actors (Hutchins, 1995a), then Integrated Operations will obviously make a difference because it introduces new actors through which cognition might take place. *What* difference Integrated Operations will make however is not given.

10.4.1 Information Overload does not Prevent Data Inaccessibility

A quick review of the quotes regarding data quantity and quality (Section 10.3.1) can lead to confusion. Some quotes show that sufficient information is available, even abundant, but that it is inconsistent and difficult to access in a timely and orderly way. Other quotes point to the opposite; that the problem is not to get hold of sufficient information, but that enough information is simply not produced. Most elaborations on the data quality describe it as poor, which is said to be a big challenge when the data are combined in large models. At the same time, better automated systems that combine large amounts of data in real-time calculations are seen as necessary. Do these statements simply display different points of view (too much and too little data), do they reveal insolvable paradoxes (combination of data is risky, and still more efficient combination of data is needed), or do they simply suggest that the linguistic construction of dichotomies and paradoxes do not do justice to the unstructured reality?

Looking at the interviews in more detail, it becomes clear that they do not reveal any real disagreement with respect to the quantity of data. There is, for example, no disagreement about the scarcity of mud data, the problem with multiple, different values for identical parameters or the general problem of handling the large amounts of data in the management of complex processes. Thus there is both a problem of too much and too little data in the industry. This is only paradoxical as long as the problems are not contextualised by having them refer to *specific* drilling contexts.

This could be illustrated by two examples: in operations where the drilling margins are especially narrow, success depends on continuous control of the hydraulic conditions in the well. Thus, the mud parameters become more important than ever. In this context, the mud data measurement regime of traditional operations is inadequate for safe and efficient operations, and most would probably agree that 'the problem for the drilling industry is that there is too *little* data'. On the other hand, the adoption of wired drill pipe is very slow in the industry, partly due to the fact that the rig teams do not have the time or resources to handle the large amounts of data this technology will provide. To them, the increased flow of data is perceived as a disadvantage rather than an advantage because they are unable to utilise the data; it would be noise rather than information. In such a context, they would probably agree that 'the problem for the drilling industry is that there is too *much* data'.

In other words, the perception of data quantity is relative to the purpose the data serves and the ability to process it.

10.4.2 The Choice is not Between Centralisation or Decentralisation

One of the key issues of disagreement between NAT (Perrow, 1984) and HRO (Weick, 1987; Weick and Roberts, 1993) is the view of the possibility to switch flexibly between a centralised and a decentralised mode of operation. In this study, this dichotomy is not as well-suited to describe the coordination modes as are the three strategies of coordination outlined by Mintzberg (1979): mutual adjustment, direct supervision and standardisation. Still, as the quotes in Section 10.3.2 show, none of them alone have the resources to account fully for the coordination.

Although the rig team's responsibility for the coordination through direct supervision of the operations is formally unambiguous, they lack the time and resources necessary to cultivate such a strategy. Thus the daily operations are, to a certain degree, coordinated by mutual adjustments between the actors at the sharp end, something which is not disavowed by an onshore rig team member stating that 'It is the offshore personnel who run the daily operations. [They] know where the shoe pinches, and they are best positioned to make the right decisions.'

This blend of direct supervision and mutual adjustment is accompanied by a strategy of standardisation. Standardisation in this context is achieved through technical and human expert systems – a core element in Integrated Operations. Both the service companies and the oil companies have expert centres that are manned and equipped to solve problems that are too complex for the offshore teams as well as the onshore rig teams (see e.g. Løwén et al., 2009). This expertise is partly based on the processing of large amounts of data in tailor-made software systems to optimise the drilling process (service company expert centre), partly on experienced personnel with access to data from a range of historical and temporary drilling operations, with which they can compare an ongoing operation to sustain or restore well control (oil company expert centre). However, neither of them have the necessary in-depth insight into each specific operation to take over the responsibility from the rig teams.

Neither centralisation nor decentralisation seem to be precise descriptions of the way drilling operations are administered. Rather, the operations are run by a combination of mutual adjustment, direct supervision and standardisation. None of these coordination strategies alone seems to be supported by adequate resources in terms of time, tools and overview. Neither does the sum of the strategies enable the different stakeholders in different positions of the operations to comfortably combine the right information to produce safe and efficient operations. This problem of coordination is a relevant issue in connection with the introduction of IO tools that continuously produce interpretations and diagnoses by combining and processing large amounts of data. As much contemporary research on organisational coordination indicates, the structural design of the organisation may be of less importance than situated and contextual action (Okhuysen and Bechky, 2009; Suchman, 1987).

10.4.3 Paperwork is an Important Ingredient in Engineering Work

Many of the drilling engineers in the study described their work as consisting of much administration and paperwork and little engineering. This was generally described as a hindrance to optimal performance. Without doubting the validity of this description or their judgement of its consequence, it is interesting to elaborate on the concept of engineering and to see what makes it so different from paperwork.

Latour and Woolgar's (1986) description of the work at The Salk Institute for Biological Studies shows that a large proportion of the biologists' work consists of meetings, report writing and paper shuffling. However, the authors do not suggest that the paperwork involved in the production and circulation of inscriptions is less 'biological' than the smaller proportion of work that the researchers spend doing biological experiments at the laboratory bench. The large amount of paperwork encountered by the drilling engineers in this study is comparable to that of the biological researchers.

This is the context into which the IO tools will be introduced. The informants' expectations for such tools include not only better and more reliable calculations. The task of distributing existing data and making them available to all stakeholders is another expectation, and access to visualisations that support discussions across disciplines and distances is yet another. Such expectations may indicate that the tools could obtain the characteristics of *boundary objects* (Star and Griesemer, 1989: 393). Thus, the expectations for the tools to support work that could be described as administrative are just as pronounced as the expectations for the tools to empower classical 'engineering' work. More Integrated Operations will produce more inscriptions, and the mobility and circulation of the inscriptions will increase. On one hand, the engineers want to spend less time on administrative work and more time on engineering work. On the other hand, what they are demanding are tools that perform both calculations and paperwork, in practice changing the role of the engineers to become even more administrative than before. Rather than inferring that the informants are inconsequent in their descriptions and that the relation between paperwork and engineering work is paradoxical, we could infer that *the distinction between engineering and paperwork is spurious*. This conclusion is in line with Hutchins' (1995b) observations in his study of distributed cognitive work. The type of interpretative work performed by petroleum engineers fits well into Hutchins' (1995a) definition of cognition in terms of coordination of representational media, papers and reports notwithstanding.

The etymological distinction between paperwork and engineering work has implications beyond the linguistic. First, the effect of the tools depends on which functions of the tools are actually acknowledged and based on this acknowledgement, who is given access to the tools. If they are thought of as black boxes that are fed with data to produce certain answers, important features of the tools could be overlooked and hence left unexploited. Second, to the extent that the introduction of the tools are accompanied by work process changes, acknowledging that engineering work cannot easily be separated from paperwork will be critical for an optimal shaping of the division of labour.

10.4.4 More Standardisation does not Necessarily Imply Less Uniqueness

According to some informants, engineering has special qualities that not only distinguish it from paperwork, but also from the domain of experience. The difference, it is stated, is that while experience is inaccurate and unreliable, the engineering methods of instrumented processes are unbiased, they take into consideration all the information available and they produce solutions based on calculations, rather than interpretations based on experience. The first quote in Section 10.3.4 exemplifies this view, which is contrasted with other statements suggesting that there are some distinct qualities connected to experience that are impossible to instrument. It is striking how these two points of view relate to each other as opposites, serving to argue respectively for and against more instrumentation of the same work processes. The first view promotes standardisation, and presupposes that this can only be obtained by instrumental aids. The latter promotes uniqueness based on experience, and connects this to human cognition.

A third view on this issue is uttered by an informant elaborating on Beta-Drill. Instead of underscoring the differences between instrumented and human-based cognition, the informant erases this distinction by describing how the tool is working; rather than focusing on generic differences between what can be achieved by technology and what can be achieved by human minds, he shows how cognition is distributed across heterogeneous actors. The tool contributes to establishing a 'connection between real-time data and historical data so that [the drilling engineers] can use historical data to recognise a situation from the real-time data, and use the experience from historical situations to solve present problems'. This process involves collecting historical drilling data, identifying, describing and categorising significant events, organising this information in a searchable database, interpreting the data flow from ongoing operations as 'fingerprints', comparing this real-time fingerprint with the fingerprints of the historical operations and using the experience from historical solutions to comparable problems. This work resembles Hutchins' (1995b) description of cognitive work.

The uniqueness, the experience and the standardisation cannot be ascribed to humans or machines alone because the human experience depends on technology, and standardisation depends on human interpretation. Hence, as the last quote in Section 10.3.4 also indicates, more standardisation does not necessarily imply less unique interpretations.

When opened up and related to practical operations, the controversy of standardisation and uniqueness appears not so controversial after all. As a descriptive theme it supplements the description of the context into which integrated drilling tools are introduced. An analysis of what influence such tools will have on the control of the operations will benefit strongly from an understanding of this context.

10.5 Redistributing Cognition with Integrated Operations

The controversies of drilling operations find resonance in the fields of safety and cognition. In the human factors tradition, both information overload and data deficiency is recognised as a challenge to human perception and cognition (see e.g. Reason, 1987; Weick and Sutcliffe, 2008). The theme of centralised/decentralised control in complex organisations is central to the two prominent safety approaches of NAT and HRO (La Porte and Consolini, 1991; Perrow, 1984). The theme of standardised and unique interpretation of data can be related to many of the works in the field of science and technology studies (e.g. Bowker and Star, 2000; Latour, 1987, 2004a). Science and technology studies also offer reflections and studies on the engineering/paperwork theme, as does the field of distributed cognition (Hutchins, 1995a, 1995b).

How IO tools will influence the safety of drilling operations is not solely a matter of the properties of the tools alone. It is argued here that it is also a matter of the context into which they are introduced. In addition, it depends on how the change process is approached. A reflexive approach would involve two considerations: the first is how one chooses to connect the controversies to the cognitive process of controlling the operations. Basing it on the technological determinism of NAT will, for example, give a different result than basing it on the more social constructivist HRO perspective. The second consideration is what role the IO tools should be granted. This should be based on how they are believed to impact the controversies.

This study indicates that the IO change process will benefit from letting the first consideration support the perspective that knowledge and cognition are created and distributed through a process involving both humans and non-humans, as supported by Hutchins (1995a) and Latour (1987). Cognitive control is dependent on the 'propagation of representations through various media' (Hutchins, 1995a: 316) and the circulation of representations between the periphery and centres of calculation (Latour, 1986, 1987). The effect of the integration of the IO tools, in turn, depends on how the integration is adapted to the controversies. Below, some questions that could be helpful in relating the IO tools to the controversies are suggested.

The controversy of *information overload* is not only about the quantity of data. As we have seen, in some parts of the operations there is a severe *data deficiency*; the availability and quality of data was also pointed to as a bottleneck. Thus we have to look beyond the mere quantity of data as the main problem. Information overload is a result of the system's cognitive abilities to process the information and could therefore, in some situations, arise also from very *little* data. The IO tools have functionality that can influence the production, distribution and processing of data. The active shaping of this influence would thus involve questions such as: who should have access to the tools, how should the tools treat different values for the same parameter, how should acceptable data quality be assured and how should data and simulations be visualised?

The controversy of *standardised versus unique interpretation of data* changes character when humans and non-humans are treated as symmetrical actors in a system of distributed cognition. One potential of the IO tools is their ability to produce representations that are immutable and combinable, thus rendering circulation between different actors within the organisation possible. Whether this form of standardisation is produced at the cost of uniqueness or not is a question of the resolution and accuracy of the interpretation, not whether it is done by a human or an instrument. The questions that should be asked thus include how fine-meshed the models should be, how accurate and reliable the sensors need to be and to what degree automatically produced interpretations are transparent and may be questioned and reinterpreted by other actors.

How the IO tools relate to the controversy of *centralisation and decentralisation* is not intrinsically enshrined in the tools themselves. One possible strategy is to integrate the tools with existing technologies and work processes in such a way that the choice between *either* centralised *or* decentralised interpretation work and control is rendered superfluous. Latour (1986, 1987) has shown how *circulation of representations* between the periphery and *centres of calculation* is a central process in cognitive work, and with the IO tools constituting a potential core element of a centre of calculation, such centres can be multiplied and distributed to a range of locations. This could render possible a decentralisation of previous typically centralised work. Thus, the allocation of interpretation and decision-making is no longer a question of availability of data, but a question of politics; who should get access to the IO tools and to what extent should the new distribution of cognition and knowledge influence the division of labour and the mandate to make decisions?

The last controversy, the controversy between *engineering and paperwork*, reflects that the choice of epistemological perspective is not uncontroversial. The fact that most of the engineers complained about the large proportion of paperwork and administrative tasks should be taken seriously. However, so must the fact that this composition of engineering and administrative tasks is a composition that the drilling organisations share with many other knowledge- and technology-intensive industries. Accepting the ontological perspective suggested here implies denying the a priori distinction between engineering work and paperwork. The handling of abstracted representations of different forms is central also to the work categorised by the informants as 'engineering'. That the paperwork and the meetings handle *more* abstracted representations does not make them less 'engineering'.

10.6 Conclusion

It is argued that the impact of new IO tools on the safety of drilling operations can be inferred only with reference to the context into which they are adopted and the way in which they are integrated with existing technologies and work processes. This context is described by a set of controversies that are characteristic of the challenges of drilling operations:

- information overload versus information availability and quality
- centralised versus decentralised control
- standardised versus unique interpretation of data, and
- engineering versus paperwork.

The identification of the controversies is important for a context-sensitive adaptation of the IO tools. By perceiving the control of the drilling operations as distributed cognitive processes, the IO tools may enter into a process of *redistributing cognition* among the heterogeneous actors. This chapter has shown that such redistribution may either amplify or reduce the controversies.

The distributed nature of cognition is, with reference to the works of Hutchins (1995a) and Latour (1987), taken as a premise. One consequence of this view on cognition is that paperwork, or administrative work in general, should be acknowledged as a central ingredient of cognitive work, and therefore also as a central ingredient of the engineering work involved in controlling the drilling operations. Such an interpretation of the controversy engineering versus paperwork is a key to deploying the three other controversies as resources for the IO change process.

How this controversy is handled in the ongoing IO change process is a touchstone for how adequate the epistemological perspective on which this study and its findings rest is to the stakeholders within the drilling organisations. If the epistemological perspective is accepted, the controversies identified can contribute to a process of redistributing cognition, a process in which the IO tools may play an important role. In that case, the work processes of drilling operations could be more integrated, less controversial and hence, safer. If, on the other hand, the epistemological perspective is rejected and the controversies are not addressed in the process of adopting the IO tools, there is no guarantee that the controversies will not be reinforced instead of resolved. This could lead to more frustration and more fragmented cognitive processes.

Chapter 11
New Tools, Old Tasks:
Discussion and Conclusions

The main objective of this book has been to explore how safety in sociotechnical work can be studied, and how the introduction of new technologies influences the safety of sociotechnical operations. Throughout the book, this objective has been approached from different angles. The first step was to understand and describe the study object – the drilling organisation/operation. This is the topic of Chapter 7, which discusses the nature of sociotechnical systems[1] and how they should be comprehended and described.[2] The next step was to investigate the characteristics of informal coordination in the occurrence of unexpected events, and how its significance for operational safety may be affected by the introduction of new technologies. Section 2.1 and Chapter 8 do this through a case study that accounts for the unplanned work of getting things back on track when operations in an offshore installation take an unexpected turn. Thereafter, in Section 2.2 and Chapter 9, the role of shared understanding[3] in multidisciplinary teams is critically examined to determine what role it actually plays in drilling operations and what role it might assume in future, more integrated operations. These discussions are also based on a case. Finally, Chapter 10 explores how new IO tools may affect the safety of drilling operations: at the same time it can be read as a meta-study that explores a method for prospective technology analysis.

The different analyses approach the main objective via different routes. In the following, their findings will be discussed and aligned into a final conclusion for the study.

---

1   Although the ambition is to produce generic knowledge, it is necessary to realise the limited value of generalisation in a study where almost all instances of the term sociotechnical systems refer to offshore drilling operations/organisations.

2   The use of 'should' in this text does not imply a categorical imperative. It must be read in the context of the book, where the imperatives serve the defined objective of understanding and describing sociotechnical safety of offshore drilling operations. Until further research has documented the generic value of the findings, 'should' might be replaced by 'could' outside of this context.

3   There are many other terms describing the same phenomenon, e.g. common operational picture and shared situation awareness.

11.1 Conceptual Descriptions Make a Difference

Several models for accidents and safety have been reviewed. While each of these models defines important mechanisms in the development of accidents and for their prevention, none of them are able to fully account for the genealogy of accidents and the recipe for safe systems. This statement is *not* a critique of those theories. A theory that describes how to design and manage a perfectly safe system will probably never be formulated since for a system to produce something, some sort of action has to take place. And by definition, action is inextricably associated with mediation and uncertainty (Latour, 2005). Rather than criticising these frameworks, this study aims at adding value to this body of literature by introducing new perspectives and new empirical material to complement the existing safety research and push the frontier of knowledge a small step further.

Normal Accident Theory (NAT) is perhaps the approach that most clearly states that it is a *theory*. It is stated in the name itself, and clearly expressed by Perrow: 'NAT is a theory of major system failures and system damage' (Perrow, 1999: 391). The simple and easily-understood 2×2 matrix of tight couplings and interactive complexity even makes it very applicable. However, the theory is at odds with some important findings in this book. In NAT the notion of complexity is used in a largely static and objectivist manner. By paying more attention to the knowledge the actors have of the system and the difference between the 'open' and the 'closed' mode of the system, a more realistic definition of complexity could be obtained. This study argues for replacing the ontological notion of complexity with an epistemological notion.[4] Likewise, the relatively one-sided focus of NAT accounts asymmetrically for the contributors to the course of action in sociotechnical systems, rendering the theory somewhat technology deterministic (see Hopkins, 2001). A less reductionist understanding of complexity could ensure that relevant actors or conditions are not excluded from the analysis. Chapters 7 and 8 deal with these issues and could be read as suggestions for theory improvement.

It is important to underscore, however, that these improvements do not simply involve adding a social component to the technical components. Although the HRO research is pursuing 'a theory of organizational behaviour under very trying conditions' (La Porte and Rochlin, 1994: 221) and is considered as *complementary* to the work of Perrow (1999), the strategy of complementing NAT by evaluating sociotechnical systems *also* in the light of HRO research to account for the social construction of safe operations (Rochlin, 1999) would lead attention away from the main argument in Chapter 7: that it is the *relations* between the entities of the system – irrespective of categories of human, technology or organisation – that are the basic entities of sociotechnical systems. Thus, it is not the isolated qualities of either the technical or the social that should be scrutinised, but the changes induced by the relations between all relevant actors. These relations – or *translations* –

---

4   Hollnagel indicates a similar preference when replacing ontological *complexity* with epistemological *intractability* (Hollnagel, 2008a).

mediate the actors and are likely to displace the frontline between programmes and antiprogrammes. Remember the case in Chapter 7: at a specific and purely technical level, the programme for a safe well intervention did not have sufficient weight, because central technical well barriers were not intact and in place;[5] at a higher and more sociotechnical level, the technical degradation of the platform actually *added* weight to the programme by shaping the rig crew's competences and their ability to cope with contingencies. This description also supports the point that is made other places in this book, that the way relations mediate the involved actors generally, and the sociotechnical processes of a drilling operation in particular, are marked by uncertainty.

This approach, which fits Callon's (1986) description of a *sociology of translation*, presupposes a principle of generalised symmetry that forbids any a priori discrimination between humans and non-humans in the descriptions of the actor-networks. Rather than reflecting any ethical programme or any perspective on the intentionality of objects, it is used here as a method motivated by a wish to account for the translations with an '[increased] level of detail and precision' (Hepsø, 2009: 39) acknowledging that 'the ingredients of controversies are a mixture of considerations concerning both Society and Nature' (Callon, 1986: 200).

11.2 Automating Interpretive Work May Lead to Unintended Consequences

One rationale for Integrated Operations is to streamline and standardise the work processes to make them safer and more efficient. Many of the informants in the study emphasised the approximate nature of today's operations and the dominant role of 'gut feeling' associated with interpretation and decision-making. Tools for monitoring, diagnosing and automating the drilling operations are among the resources Integrated Operations may mobilise to respond to such descriptions. However, to understand which role such tools may play in the operations and what the consequences of substituting instrumentation, calculation and visualisations for interpretive work may be, it is necessary to explore this interpretive work and to understand what the contribution of interpretive work is to the safety of the operations. This is primarily explored in Chapter 8, but Chapter 9 also deals with the theme.

Articulation work can be characterised as the 'work that gets things back "on track" in the face of the unexpected, and modifies action to accommodate unanticipated contingencies' (Star and Strauss, 1999: 10) and 'work that manages the [consequences] of [the] distributed aspect of the work' (Bowker and Star, 2000:

---

5   Cf. Section 5.5 and the need to support a programme with material and immaterial loads. In the case of the hotel manager (Latour, 1991) these included oral messages, written notes and a metal weight attached to the key. In the case of the drilling operations, the loads that might have been necessary for the drilling programme to succeed were the technical barriers that were not intact and in place.

310). This type of work is described in detail in the case study of Section 2.1 and Chapter 8. The most obvious point is that articulation work fills a gap that is not covered by formal work descriptions but still is necessary to 'make work work' (Schmidt, 2010: 184). Another effect of this articulation work that also deserves attention is the effect of articulation work in making invisible work visible. Not only is the articulation work itself 'invisible to rationalized models of work' (Star and Strauss, 1999: 10), implicating that articulation work must be explicitly described and presented for its value to be acknowledged. Also the different primary work processes that are articulated and the situated interconnectedness of these work processes are to a large extent invisible to others than those who execute them, and the articulation work also makes work visible to the rest of the organisation, thus contributing to the organisation's system comprehension.

Since this non-standardised work is of such importance for managing the consequences of division of labour, it may seem a paradox that new tools and technologies that aim at *standardising* work processes are introduced to better handle the interconnectedness of different actors and work processes. However, the boundary between primary work and articulation work is not necessarily unbridgeable and the lifting of articulation work into the domain of primary work should not be excluded as a possibility. Such a substitution does, however, presuppose an explicit acknowledgement of articulation work and its role in the operations.

11.3 Interpretive Work in a Context of Multiple Goals

Shared understanding is intended to assume a central role in Integrated Operations.[6] The nearly unambiguous expectations towards the concept in the industry and among IO researchers[7] to support interpretive work almost beg for a curious and critical examination. In Section 2.2 and Chapter 9, the interpretation of data in a multidisciplinary team is reviewed in a setting of discrepant depth measurements to explore the role of shared understanding for safe and efficient operations in a sociotechnical system. The main conclusion to this case is, somewhat surprisingly, that the promise of the concept is *not* justified by empirical investigation, and that it is *not* sufficiently empirically nor theoretically grounded that it plays a central role in the production of safe or unsafe outcomes of operations.

It is necessary to underscore that the case study does not say anything about the ability of any IO tool to produce or support shared understanding; actually it reviews a case where such IO technologies are not being actively used at all. Although some may find it tempting, the case study should not be read as an

---

6   Explicitly expressed by for example Rommetveit et al. (2008b) as an argument for IO technology intended to produce a common operational picture.

7   The only critical comment I have registered by IO stakeholders is the suggestion that shared understanding could lead to groupthink.

argument for a need for better tools to create and support this shared understanding. What is examined are the ontological conditions necessary for shared situational understanding to have a positive impact *at all*. In drilling operations, given the different goals of the involved actors, these conditions are not found to be present. Thus the scarce resource is not representational technologies, but rather a common goal among the stakeholders and a strategy for how it could be pursued in the case of disagreement between the professional domains.

The expectations of shared understanding are not fulfilled in a context of different goals. In the absence of an unambiguous common goal to provide the multidisciplinary team a means for directing and coordinating the work and the decisions, there is, however, another type of entity that has proved useful in other contexts, namely the *boundary object*. Star and Griesemer (1989) first wrote about boundary objects in connection with a study of the multidisciplinary work at the Museum of Vertebrate Zoology at the University of California. The collaborating conditions in the drilling organisations could be compared with those in Griesemer and Star's case study; different groups of experts from different social worlds, with different perspectives and goals. Star and Griesemer labelled the coordinating mechanisms that made collaboration between the different groups possible *boundary objects*. Boundary objects are

> objects which are both plastic enough to adapt to local needs and the constraints of the several parties employing them, yet robust enough to remain a common identity across sites. They are weakly structured in common use, and become strongly structured in individual use. These objects may be abstract or concrete. (Star and Griesemer, 1989: 393)

Shared understanding is expected by many to play a role as a coordinating mechanism in future Integrated Operations. A question is, however, whether shared understanding as such, or the IO technologies that are to provide this shared understanding, may function as such a boundary object in the interpretation work and the decision-making processes. The answer offered by this book is 'no'.

The shared understanding supported by the technologies for monitoring and diagnosing technologies differs from boundary objects in a crucial way: the intention with these tools is to represent the 'true state' of the well and the drilling process, an unambiguous representation of the state of affairs based on mathematical models and real-time sensor data. Also, these representations are meant to align the understandings across disciplines and sites. This is in great contrast to the 'plastic' nature of Star and Griesemer's boundary objects, and their 'weak structure' in common use and 'strong structure' in individual use. While shared understanding points towards *consensus* as a prerequisite for cooperative work, boundary objects do not:

> Common myths characterize scientific cooperation as deriving from a consensus imposed by nature. But if we examine the actual work organization of scientific

enterprises, we find no such consensus. Instead, we find that scientific work neither loses its internal diversity nor is consequently retarded by lack of consensus. Consensus is not necessary for cooperation nor for the successful conduct of work. (Star and Griesemer, 1989: 388)

In fact, the ambiguity is a necessity for the boundary objects to work:

It is unrealistic and counter-productive to try to destroy all uncertainty and ambiguity in boundary objects. By their very nature, boundary objects need appropriate degrees of both in order to work. (Bowker and Star, 1991: 79)

This should have consequences for reliance on IO technologies that are designed to produce or support shared understanding. If there is a need for better communication across different locations and professions, tools and work processes may surely support this. However, this need should not be confused with a need for shared understanding.

11.4 Identifying Matters of Concern

A challenge of evaluating the consequences of the ongoing change process of IO is that one to a large extent deals with visions and prospective technologies. Although IO is an ongoing process and much of the already existing technologies and work processes of drilling operations fit well into the IO definition, a large portion of the research on IO and safety still relate to parts of IO yet to be implemented or developed. What separates the approach of this study from much of the existing research on IO and safety is the strong focus on existing operations and their characteristics, their constraints and their strengths. This is not as much a consequence of definitions of what should or should not be interpreted as belonging to the IO discourse as it is of the epistemological perspective that when new tools and work processes are introduced to an existing context, one cannot understand the consequences by looking at the *new* elements alone. The *old*, the existing operations, have developed into a strong and stable work-net (Latour, 2005: 143) through a long period of time, thus the question of how the *old* will change the *new* could be just as relevant as how the *new* will change the *old*.

A major strategy in this study has therefore been to conceptualise and elaborate on the work and challenges in drilling operations *as such*, integrated or not, in order to establish an IO-relevant understanding of the operations into which new tools and work processes will be – or have already been – introduced. Some of the consequences of such descriptions have been elaborated on previously: first, descriptions of the design and characteristics of new tools should not be used unilaterally to account for change processes and their impact on sociotechnical systems. Second, the descriptions of the complexity of a sociotechnical system, and the impact on complexity of new technologies and work processes, could benefit from an epistemological approach. Such an approach regards complexity as a transient entity, and it renders it a less relevant indicator for system safety than the approach of for example NAT (Perrow, 1999) does. Third, nuances of the work of collaboration and coordination have been illustrated with respect to the difference between primary work and articulation work, and the opportunities and constraints associated with these different types of work. Whereas primary work is characterised by visibility, planning and predictability and thus can be straightforwardly addressed by the streamlining and automation aspects of IO, articulation work is less visible, and it is associated with contributions to awareness and increased knowledge in the sociotechnical system that may not be easily achieved by the tools and work processes of IO. Fourth, the preconditions for interpretation of data in multidisciplinary teams in IO are shown to be less dependent on shared situational understanding[8] than generally believed.

All of these four findings are closely associated with and point towards the fifth finding of the study: the way in which new technologies and work processes will influence the safety of drilling operations is marked by uncertainty, and a fruitful way of approaching and influencing this uncertainty is to map the dominant controversies within the organisation/operations and to address them explicitly in the IO change process. Being core characteristics of the existing drilling operations, the identification of the importance of these controversies strengthen this book's recurring point, which deserves the status as the general conclusion of the book: *a sociotechnical approach to evaluate the influence of Integrated Operations on the safety of drilling operations consistently leads attention to the existing operations. The tools may be new, but the tasks are old. The preferred method of inquiry is thus to investigate existing practices in the presence of specific IO technologies and work processes.*

11.5 Concluding Remarks

The main objective of this study involves two dimensions. The first dimension is how safety in sociotechnical work can be studied. The second is how the introduction of new technologies will influence the safety of sociotechnical operations. Although a general conclusion of the study is offered above, a recapitulation of the theoretical frameworks and methodological themes reviewed in Chapter 5 and Chapter 6 may hint at an alternative way of framing the answer to the questions.

Regarding the first question, these chapters presents what may be understood as four different approaches to understanding safety – cf. Figure 6.1. The approach elaborated upon in this study makes use of ethnographic methods to explore the heterogeneous and relational (Latour, 1996a) nature of sociotechnical work and its

---

8   Shared understanding is used as a collective term for a common operational picture, shared situational awareness, shared mental models, etc.

outcomes. The epistemological premises for this approach have been elaborated in Section 5.5.[9] What has been systematically shown throughout this study is that *by adopting those epistemological premises and investigating the sociotechnical work and assets without assuming in advance the factors or actors that qualify for making a difference, novel insights about how safety is affected may be obtained. An important point is that these insights may be difficult to obtain and justify by the use of generic models and conceptualisations of safety and accident genealogies that are offered by much of the existing safety research and literature.*

Consequently, the second question must be approached indirectly because the chosen methodology does not involve predictive models. Thus, *it is the not the direct consequences of IO, but rather the conditions that will affect and be affected by IO that may and should be identified and elaborated on.* Whereas this study has downplayed the role of shared understanding as such a condition,[10] existing controversies in drilling operations are identified as a significant area of concern.[11]

11.6 Closing the Methodological Loop

Why look at existing operations when the main objective is to study the effects of the Integrated Operations of the future? The arguments above can be further elaborated by revisiting the methodological basis for the study. Being largely inspired by writings in the fields of Actor-Network Theory and science and technology studies, the resonance between the main conclusion and these writings, whose general methodological characteristics relevant to this study are reviewed in Section 5.5 and Chapter 6, is not surprising. At this point, it could be useful to revisit the methodological approach in a new light to see whether this resonance may still support the conclusion.

The approach of the study has been based on the strategy of, metaphorically speaking, following the ball and not the player. The ball being a metaphor for the object of desire, namely continuous organisational control over the drilling operations, IO technologies and work processes have only been taken into account when they have come explicitly into play. This approach also reflects a *flat earth* as described by Latour (1996b),[12] whose description does not discriminate between interaction and context or between agencies and structures. Thus, Integrated Operations is not treated as a new *underlying* context or structure in which another type of work will take place, nor are traditional drilling operations treated as an *underlying* context or structure into which new IO technologies and work processes are simply immersed. However, the empirical investigations indeed reveal several

---

9   Some key words may be relationism, heterogeneity and symmetry.
10 Cf. Section 2.2 and Chapter 9.
11 Cf. Chapter 10.
12 See also Section 5.5.10.

aspects of 'traditional operations' that are still likely to play an important role, although Integrated Operations introduce new dimensions to the operations.

The question is not what is new and what is old; the question is which entities should enter into the descriptions of the drilling operations in order for those descriptions to be representative, relevant and useful? The answer proposed in this study, based on empirical investigations, is that it is fruitful to let the descriptions take the existing practices as a point of departure and seek to identify the areas that will affect, and be affected by, the new technologies and work processes.

This conclusion may seem tautological, having already said that its supporting arguments themselves are influenced by such an epistemological point of departure of Actor-Network Theory. However, the tautology is merely apparent. Indeed, the value of the conclusion depends on the arguments that lead up to it, and had it been the case that those arguments merely 'transported' the methodological framework through the study and towards the conclusion, the conclusion would be redundant. However, it is this author's opinion that the methodological framework's role in the production of novel insights should be that of a catalyst: it supports the process, but *it does not itself enter into it*. When a conclusion that is well-grounded in empirical investigations can be read as an argument for a methodology, this argument is not necessarily tautological.

The answer to the question of how to evaluate the value and independency of such a conclusion is that it should be recognised from the empirical investigation and findings. The findings presented in the book are not merely vehicles for the methodological framework. They are novel insights that were not postulated or foreseen in advance, and having been constructed they are no longer dependent on that framework. The best evidence for that is the recognition of the findings from the informants. The fact that the conclusion in this case is strongly coherent with the methodological point of departure should therefore not be interpreted as a tautology, but rather serve as a confirmation of the *adequacy* of the framework and the strength of the conclusion.

11.7 Recommendations for Further Research

In accordance with Weick's (1995) view of theory, this study constitutes a theory of safety in Integrated Operations. At the continuum of theory it is, however, at the stage of 'interim struggle' (Weick, 1995: 385) rather than of 'strong theory'. The following three points indicate how the theoretical contributions of this book may be supplemented and strengthened.

11.7.1 Further Empirical Investigation

Empirical investigations constitute the backbone of this study. The methodological approach clearly favours empirical exploration to evaluation based on predefined generic theoretical models. However, the empirical study does not pretend to cover all aspects of integrated drilling operations. Largely limited to tools for monitoring, diagnosis and automation, many important aspects of Integrated Operations have been left out. The reason for this is obviously that an all-embracing study would be an impossible task given the available resources. This study should thus be considered as a small contribution to a more comprehensive, incremental process. Future research should be able to produce both specific and generic knowledge on the consequences of a wide range of technologies and work processes that are likely to be introduced to the industry in the coming years. As a starting point, it would be useful to undertake a comprehensive mapping of both implemented and planned IO initiatives. This would serve as a point of departure for the next points described below.

11.7.2 Further Developments of a Methodological Framework for Integrated Operations and Safety

In order to proceed from the 'interim struggles' of this study to a more comprehensive and applicable theory, the methodological framework should be deepened, widened and consolidated. The development of models that could guide the evaluation process would represent a useful contribution. Such models should both reflect a theoretical foundation for the relationship between Integrated Operations and safety, and provide a generic means for evaluating the safety consequences of new IO initiatives.

This is obviously a very ambitious recommendation; the aim for a generic model that may be used for evaluating specific IO tools and work processes may resemble a delusion of grandeur. The model should not, however, cover areas that have not been covered by empirical investigations. The type of generic model proposed by Perrow to categorise industries with respect to interactive complexity and tightness of couplings is beyond of the scope of this recommendation. The proposition is rather that the model should function as a methodological recommendation for the investigation of specific IO change processes.

11.7.3 Develop Recommendations for Design, Implementation and Use

A third challenge for future research would be to develop recommendations for the design, implementation and use of IO technologies and work processes, and their integration with existing practices. Such recommendations will, if grounded in a thorough elaboration of the first two points, represent an important contribution to a sensitive and safe IO development in all areas of the IO change process.

References

Aeronautica Civil de Colombia 1996. *Aircraft Accident Report: Controlled Flight into Terrain, American Airlines Flight 965, Boeing 757-223, N651AA near Cali, Colombia, 20 December 1995*. Bogota, Colombia, Aeronautica Civil.

Albrechtsen, E. ed. 2010. *Essays on Socio-Technical Vulnerabilities and Strategies of Control in Integrated Operations*. Mines Paris Tech/SINTEF/NTNU, Trondheim.

Albrechtsen, E. and Besnard, D. (2013), *Oil and Gas, Technology and Humans: Assessing the Human Factors of Technological Change*. Farnham: Ashgate.

Almklov, P.G. 2006. *Kunnskap, Kommunikasjon og Ekspertise. Et Antropologisk Studium av En Tverrfaglig Ekspertgruppe i Oljeindustrien [Knowledge, communication and expertise]. Trondheim*. Norway: Norges Teknisk-Naturvitenskapelige Universitet, Trondheim: Fakultet for Samfunnsvitenskap og Teknologiledelse.

Almklov, P.G. 2008. Standardized data and singular situations. *Social Studies of Science*, 38(6), 873–897.

Almklov, P.G. and Hepsø, V. 2011. Between and beyond data: how analogue field experience informs the interpretation of remote data sources in petroleum reservoir geology. *Social Studies of Science*, 41(4), 539–561.

Almklov, PG, Østerlie, T and Haavik, T.K. 2012. Punctuation and Extrapolation: Representing a Subsurface Oil Reservoir. *Journal of Experimental & Theoretical Artificial Intelligence*, 24(3), 329–350.

Andresen, G., Grøtan, T.O., Johnsen, S.O., Rosness, R., Sivertsen, T., Steiro, T., Thunem, A. and Tveiten, C. 2006. *Samhandling over Avstand – Erfaringer av Relevans for Petroleumsbransjen [Collaboration Across Distance – Relevant Experiences from the Petroleum Industry]*. Trondheim: SINTEF.

Antonsen, S. 2009a. Safety culture assessment: a mission impossible? *Journal of Contingencies and Crisis Management*, 17(4), 242–254.

Antonsen, S. 2009b. *Safety Culture: Theory, Method and Improvement*. Farnham: Ashgate Publishing.

Artman, H. and Garbis, C. 1998. Situation Awareness as Distributed Cognition. In Cognition and Cooperation. Proceedings of 9th Conference of Cognitive Ergonomics, Green, T., Bannon, L., Warren, C., and Buckley, J. eds, pp. 151–156 (Republic of Ireland: Limerick).

Bannon, L. 2000. Understanding Common Information Spaces in CSCW. Paper to the Workshop on Cooperative Organisation of Common Information Spaces, August 23–25 2000, Technical University of Denmark.

Bannon, L. and Bødker, S. 1997. In J.A. Hughes, W. Prinz, T.A. Rodden and K. Schmidt (eds): ECSCW'97: Proceedings of the Fifth European Conference on Computer-Supported Cooperative Work, 7–11 September 1997, Lancaster, U.K. Dordrecht: Kluwer Academic Publishers.

Bateson, G. 1982. Difference, double description and the interactive designation of self. In *Studies in Symbolism and Cultural Communication*, edited by Hanson, F. A., Lawrence, KS: University of Kansas.

Bateson, G. 2000. *Steps to an Ecology of Mind*. Chicago, IL: University of Chicago Press.

Beck, U. 1992. *Risk Society: Towards a New Modernity*. London: Sage.

Beers, P.J., Boshuizen, H.P.A., Kirschner, P.A. and Gijselaers, W.H. 2005. Computer support for knowledge construction in collaborative learning environments. *Computers in Human Behavior*, 21(4), 623–643.

Bijker, W.E. 1995. *Of Bicycles, Bakelites, and Bulbs: Toward a Theory of Sociotechnical Change*. Cambridge, MA: MIT Press.

Bijker, W.E., Hughes, T.P. and Pinch, T. 1987. *The Social Construction of Technological Systems: New Directions in the Sociology and History of Technology*. Cambridge, MA: MIT Press.

Billings, C.E. 1996. Situation awareness measurement and analysis: a commentary, in *Experimental Analysis and Measurement of Situation Awareness*, edited by Garland, D.J. and Endsley, M.R. Daytona Beach, FL: Embry-Riddle Aeronautical University Press, 1–6.

Bossen, C. 2002. The Parameters of Common Information Spaces: The Heterogeneity of Cooperative Work at a Hospital Ward. Paper to the 2002 ACM conference on Computer-Supported Cooperative Work, 16–20 November, New Orleans, LA, USA.

Bowker, G. and Star, S.L. 1991. Situations Vs. Standards in Long-Term, Wide-Scale Decision-Making: The Case of the International Classification of Diseases. Paper to the Twenty-Fourth Annual Hawaii International Conference on System Sciences, Kauai, Hawaii, 7–11 Jan 1991.

Bowker, G.C. and Star, S.L. 2000. *Sorting Things Out: Classification and its Consequences*. Cambridge, MA: MIT Press.

Brattbakk, M., Østvold, L.-Ø. and Hallvard Hiim, C. 2005. *Gransking av Gassutblåsning på Snorre A, Brønn 34/7-P31 A 28.11.2004. [Investigation of the Gas Blowout at Snorre A, Well 34/7-P31 a 28.11.2004]*. Stavanger: Ptil (Petroleum Safety Authority Norway).

Bremdal, B and Korsvold, T. 2012. Knowledge Markets and Collective Learning: Designing Hybrid Arenas for Learning Oriented Collaboration, in Rosendahl, T. and Hepsø, V. 2013. *Integrated Operations in the Oil and Gas Industry: Sustainability and Capability Development.* Hershey, PA: IGI Global.

Brown, J.S. 1991. Organizational learning and communities-of-practice: toward a unified view of working, learning, and innovation. *Organization Science*, 2(1), 40–57.

Callon, M. 1986. Some elements of a sociology of translation: domestication of the scallops and the fishermen of St Brieuc bay, in *Power, Action and Belief: A New Sociology of Knowledge*, edited by Law, J. London: Routledge, 196–223.

Callon, M. 1987. Society in the making: the study of technology as a tool for sociological analysis, in *The Social Construction of Technological Systems: New Directions in the Sociology and History of Technology*, edited by Bijker, W.E., Hughes, T.P. and Pinch T., Cambridge, MA: MIT Press, 83–103.

Clark, H.H. and Brennan, S.E. 1991. Grounding in communication, in *Perspectives on Socially Shared Cognition*, edited by Resnick, L.B., Levine, J.M. and Teasley, S.D. Washington, DC: American Psychological Association, 127–149.

Dekker, S. 2006. Resilience engineering: chronicling the emergence of confused consensus, in *Resilience Engineering. Concepts and Precepts*, edited by Hollnagel, E., Woods, D.D. and Leveson, N. Aldershot: Ashgate, 77–92.

Dekker, S. and Hollnagel, E. 2004. Human factors and folk models. *Cognition, Technology & Work*, 6(2), 79–86.

Deleuze, G. 1993. *The Fold: Leibniz and the Baroque*. London: Athlone Press.

Edwards, A., Mydland, O. and Henriquez, A. 2010. The Art of Intelligent Energy (iE)-Insights and Lessons Learned from the Application of iE. Paper to the SPE Intelligent Energy Conference and Exhibition 23–25 March 2010, Utrecht, The Netherlands.

Endsley, M.R. 1988. Design and Evaluation for Situation Awareness Enhancement. In Proceedings of the Human Factors Society 32nd Annual Meeting, 24–28 October, Santa Monica, CA: Human Factors and Ergonomics Society, 97–101.

Endsley, M.R. 1999. Shared situation awareness in the flight deck-ATC system. *IEEE Aerospace and Electronic Systems Magazine*, 14(8), 25–30.

Endsley, M.R. 2000. Theoretical underpinnings of situation awareness: a critical review, in *Situation Awareness: Analysis and Measurement*, edited by Endsley, M.R. and Garland, D.J. Mahwah, NJ: Lawrence Erlbaum Associates, 3–28.

Endsley, M.R. and Jones, W.M. 1997. *Situation Awareness, Information Dominance, and Information Warfare (No. AL/CF-TR-1997-0156)*. *Wright-Patterson Air Force Base, OH: U.S. Air Force, Armstrong Laboratory.*

Engeström, Y. and Middleton, D. 1996. *Cognition and Communication at Work*. Cambridge: Cambridge University Press.

Fields, B. 2005. Representing collaborative work: the airport as common information space. *Cognition, Technology .& Work*, 7(2), 119–133.

Flach, J.M. 1995. Situation awareness: proceed with caution. *Human Factors*, 37(1), 149–157.

Garfinkel, H. 1967. *Studies in Ethnomethodology*. Englewood Cliffs, NJ: Prentice-Hall.

Garfinkel, H. and Rawls, A.W. 2002. *Ethnomethodology's Program: Working out Durkheim's Aphorism*. Lanham, MD: Rowman & Littlefield.

Gerson, E.M. and Star, S.L. 1986. Analyzing due process in the workplace. *ACM Transactions on Information Systems*, 4(3), 257–270.

Gibson, J. 1961. The contribution of experimental psychology to the formulation of the problem of safety: a brief for basic research, in *Behavioral Approaches to Accident Research. New York*: Association for the Aid of Crippled Children New York, 77–89.

Giddens, A. 1987. *Social Theory and Modern Sociology*. Stanford, CA: Stanford University Press.

Giere, R. 2002. Scientific cognition as distributed cognition, in *The Cognitive Basis of Science*, edited by Carruthers, P., Stitch, S. and Siegal, M. Cambridge: Cambridge University Press, 285–299.

Giere, R. and Moffatt, B. 2003. Distributed cognition: where the cognitive and the social merge. *Social Studies of Science*, 33(2), 1–10.

Glaser, B.G. 1994. *More Grounded Theory Methodology: A Reader*. Mill Valley, CA: Sociology Press.

Glaser, B.G. and Strauss, A.L. 1967. *The Discovery of Grounded Theory: Strategies for Qualitative Research*. Chicago, IL: Aldine.

Goodwin, C. 1994. Professional vision. *American Anthropologist*, 96(3), 606–633.

Graham, B., Reilly, W.K., Beinecke, F., Boesch, D.F., Garcia, T.D., Murray, C.A. and Ulmer, F. 2011. *Deep Water: The Gulf Oil Disaster and the Future of Offshore Drilling: Report to the President*. Washington, DC: National Commission on the BP Deepwater Horizon Oil Spill and Offshore Drilling.

Grice, H.P. 1975. Logic and conversation in *Syntax and Semantics. 3, Speech Acts*, edited by Cole, P. and Morgan, J.L. New York: Seminar Press, 225–242.

Grøtan, T.O., Albrechtsen, E., Rosness, R. and Bjerkebaek, E. 2010. The influence on organizational accident risk by integrated operations in the petroleum industry. *Safety Science Monitor*, 14(1), 1–11.

Grøtan, T.O., Albrechtsen, E. and Skarholt, K. 2009. How Shared Situational Awareness Influences Organizational Accident Risk in the Offshore Oil Industry. In Proceedings of the eighteenth European Safety and Reliability Conference (ESREL), 7–9 September, Prague: Taylor and Francic.

Haavik, T.K. 2012. Challenging controversies: a prospective analysis of the influence of new technologies on the safety of offshore drilling operations. *Journal of Contingencies and Crisis Management*, 20(2), 90–101.

Haavik, T.K. 2011. On components and relations in sociotechnical systems. *Journal of Contingencies and Crisis Management*, 19(2), 99–109.

Haavik, T.K. 2011. Chasing shared understanding in drilling operations. Cognition. *Technology & Work*, 13(4), 281–294.

Haavik, T.K. 2010. Making drilling operations visible: the role of articulation work for organisational safety. *Cognition, Technology & Work*, 12(4), 285–295.

Hackett, E.J., Amsterdamska, O., Lynch, M. and Wajcman, J. 2008. *The Handbook of Science and Technology Studies*. Cambridge, MA: MIT Press.

Haddon, W. 1980. The basic strategies for reducing damage from hazards of all kinds. *Hazard Prevention*, 16(11), 8–12.

Hampson, I. 2005. Invisible work, invisible skills: interactive customer service as articulation work. *New Technology, Work, and Employment*, 20(2), 166–181.

Hanseth, O., and Monteiro, E. Changing Irreversible Networks, in W. R. J. Baets (ed.), proceedings of the Sixth European Conference on Information Systems, 4–6 June, Aix-en-Provence, France.

Heath, C., Knoblauch, H. and Luff, P. 2000. Technology and social interaction: the emergence of 'workplace studies'. *British Journal of Sociology*, 51(2), 299–320.

Heath, C. and Luff, P. 1992. Collaboration and control: crisis management and multimedia technology in London Underground line control rooms. *Computer-Supported Cooperative Work*, 1(1), 24–48.

Heinrich, H.W. 1931. *Industrial Accident Prevention: A Scientific Approach*. New York: McGraw-Hill.

Helset, H., Halsey, G., Larsen, I., Romdhane, A. and Nielsen, J. 2011. *Investigation of Feasibility and Potential for Sub-Surface Imaging Using Wired Drillpipe in Connection with Seismic While Drilling*. Trondheim: Center for Integrated Operations in the Petroleum Industry.

Hepsø, V. 2002. *Translating and Circulating Change: The Career of an Integrated Organization and Information Technology Concept*. Trondheim: Sosialantropologisk institutt, Fakultet for samfunnsvitenskap og teknologiledelse, Norges teknisk-naturvitenskapeligeuniversitet.

Hepsø, V. 2006. When Are We Going to Address Organizational Robustness and Collaboration as Something Else Than a Residual Factor? Paper to the 2006 SPE Intelligent Energy Conference and Exhibition, 11–13 April, Netherlands, Amsterdam.

Hepsø, V. 2009. *Leading Research in Technoscience: Insider Social Science in Socio-Technological Change*. Saarbrücken: VDM Verlag Dr. Müller.

Herbert, M., John, A. and James, R. 2008. ConocoPhillips Onshore Drilling Centre in Norway – a Virtual Tour of the Centre and Offshore Operations. Paper to the Intelligent Energy Conference and Exhibition, 25–27 February 2008, Amsterdam, The Netherlands.

Hollnagel, E. 2004. *Barriers and Accident Prevention*. Aldershot: Ashgate.

Hollnagel, E. 2008a. The changing nature of risks. *HFESA Ergonomics Australia Journal*, 22(1–2), 33–46.

Hollnagel, E. 2008b. From Protection to Resilience: Changing Views on How to Achieve Safety. Proceedings of the Eighth International Symposium of the Australian Aviation Psychology Association, 8–11 April, Sydney, Australia.

Hollnagel, E. 2009. *The ETTO Principle: Efficiency-Thoroughness Trade-Off – Why Things That Go Right Sometimes Go Wrong*. Aldershot: Ashgate.

Hollnagel, E., Nemeth, C. and Dekker, S. 2008. *Resilience Engineering Perspectives: Remaining Sensitive to the Possibility of Failure*. Aldershot: Ashgate.

Hollnagel, E., Woods, D.D. and Leveson, N. eds 2006. *Resilience Engineering: Concepts and Precepts*. Aldershot: Ashgate.

Hollnagel, E. and Woods, D.D. 2005. *Joint Cognitive Systems: Foundations of Cognitive Systems Engineering*. Boca Raton, FL: Taylor & Francis/CRC Press.

Hollnagel, E., Pariès, J., Woods, D.D. and Wreathall, J. 2011. *Resilience Engineering in Practice: A Guidebook*. Farnham: Ashgate Publishing.

Hopkins, A. 2001. Was Three Mile Island a 'Normal Accident'? *Journal of Contingencies and Crisis Management*, 9(2), 65–72.

Hutchins, E. 1995a. *Cognition in the Wild*. Cambridge, MA: MIT Press.

Hutchins, E. 1995b. How a cockpit remembers its speeds. *Cognitive Science*, 19(3), 265–288.

Hyne, N.J. 2001. *Nontechnical Guide to Petroleum Geology, Exploration, Drilling, and Production*. Tulsa, OK: PennWell.

Iversen, F., Cayeux, E., Dvergsnes, E.W., Gravdal, J.E., Vefring, E.H., Mykletun, B., Torsvoll, A., Omdal, S. and Merlo, A. 2006. Monitoring and Control of Drilling Utilizing Continuously Updated Process Models. Paper to the IADC/SPE Drilling Conference, Miami, Florida, 21–23 February 2006.

Iversen, F., Cayeux, E., Dvergsnes, E.W., Welmer, M., Torsvoll, A. and Merlo, A. 2007. Demonstrating a New System for Integrated Drilling Control. Paper to the AADE National Technical Conference and Exhibition, Houston, Texas, 10–12 April, 2007.

Janis, I. 1972. *Victims of Groupthink*. Boston, MA: Houghton Mifflin.

Jernelöv, A. and Lindén, O. 1981. Ixtoc I: a case study of the world's largest oil spill. *Ambio*, 299–306.

Johnsen, S.O. 2012. *An Investigation of Resilience in Complex Socio-Technical Systems to Improve Safety and Continuity in Integrated Operations*. Trondheim: Norwegian University of Science and Technology.

Johnson, W. 1980. *Mort Safety Assurance Systems*. New York: Marcel Dekker Inc.

Kaarstad, M. and Rindahl, G. 2011. Shared collaboration surfaces to support adequate team decision processes in an integrated operations setting, in *Advances in Safety, Reliability and Risk Management*, edited by Bérenguer, Grall & Guedes Soares. London: Taylor and Francic Group.

Kaarstad, M., Rindahl, G., Torgersen, G. and Drøivoldsmo, A. 2009. Interaction and Interaction Skills in an Integrated Operations Setting. Paper to the IEA 2009, 17th World Congress on Ergonomics, Beijing, China, 9–14 August.

Kindingstad, T. 2002. *Norwegian Oil History*. Stavanger: Wigestrand.

Klein, G. 2005. Common ground and coordination in joint activity, in *Organizational Simulation*, edited by Rouse, W.B. and Boff, K.R. Hoboken, NJ: Wiley-Interscience, 139–157.

Knorr-Cetina, K. 1981. *The Manufacture of Knowledge: An Essay on the Constuctivist and Contextual Nature of Science*. Oxford: Pergamon.

Kongsvik, T. 2003. *Hvilke Barrierer? Ansattes Vurdering Av Sider Ved Sikkerhetskulturen – Snorre A* (Which Barriers? Employees' Perception of Aspects of Safety Culture – Snorre Alpha). Trondheim: Studio Apertura.

Koschmann, T. and LeBaron, C.D. 2003. Reconsidering Common Ground: Examining Clark's Contribution Theory in the Operating Room. Paper to the European Computer-Supported Cooperative Work Conference (ECSCW 03), 14–18 September, Helsinki, Finland.

La Porte, T.R. 1996. High-reliability organizations: unlikely, demanding and at risk. *Journal of Contingencies and Crisis Management*, 4(2), 60–71.

La Porte, T.R. and Consolini, P.M. 1991. Working in practice but not in theory: theoretical challenges of high-reliability organizations. *Journal of Public Administration Research and Theory*, 1(1), 19–47.

La Porte, T.R. and Rochlin, G. 1994. A rejoinder to Perrow. *Journal of Contingencies and Crisis Management*, 2(4), 221–227.

Latour, B. 1986. Visualization and cognition: thinking with eyes and hands. *Knowledge and Society: Studies in the Sociology of Culture Past and Present*, 6, 1–40.

Latour, B. 1987. *Science in Action: How to Follow Scientists and Engineers through Society*. Milton Keynes: Open University Press.

Latour, B. 1988. *The Pasteurization of France*. Cambridge, MA: Harvard University Press.

Latour, B. 1991. Technology is society made durable, in *A Sociology of Monsters: Essays on Power, Technology and Domination*, edited by Law, J. London: Routledge, 103–131.

Latour, B. 1992. Where are the missing masses? The sociology of a few mundane artifacts, in *Shaping Technology/Building Society: Studies in Sociotechnical Change*, edited by Bijker, W.E. and Law, J. Cambridge, MA: MIT Press, 225–264.

Latour, B. 1993. *We Have Never Been Modern*. New York: Harvester Wheatsheaf.

Latour, B. 1996a. *Aramis or the Love of Technology*. Cambridge, MA: Harvard University Press.

Latour, B. 1996b. Foreword: the flat-earthers of social theory, in *Accounting and Science: Natural Inquiry and Commercial Reason*, edited by Power, M. Cambridge: Cambridge Univsersity Press, xi–xvii.

Latour, B. 1996c. On Actor-Network Theory. A few clarifications plus more than a few complications Available at: http://www.cours.fse.ulaval.ca/edc-65804/latour-clarifications.pdf.

Latour, B. 1996d. Review: *Cognition in the Wild*. *Mind, Culture and Activity*, 3(1), 54–63.

Latour, B. 1999a. Circulating references. Sampling the soil in the Amazon forest, in *Pandora's Hope: Essays on the Reality of Science Studies*, edited by Latour, B. Cambridge, MA: Harvard University Press, 24–79.

Latour, B. 1999b. *Pandora's Hope: Essays on the Reality of Science Studies*. Cambridge, MA: Harvard University Press.

Latour, B. 2000. When things strike back: a possible contribution of 'science studies' to the social sciences. *The British Journal of Sociology*, 51(1), 107–123.

Latour, B. 2003a. Is re-modernization occurring – and if so, how to prove it? A commentary on Ulrich Beck. *Theory, Culture and Society*, 20(2), 35–48.

Latour, B. 2003b. The promises of constructivism, in *Chasing Technoscience: Matrix for Materiality*, edited by Ihde, D. and Selinger, E. Bloomington, IN: Indiana University Press, 27–46.

Latour, B. 2004a. How to talk about the body? The normative dimension of science studies. *Body Society*, 10(2–3), 205–229.

Latour, B. 2004b. *Politics of Nature: How to Bring the Sciences into Democracy*. Cambridge, MA: Harvard University Press.

Latour, B. 2005. *Reassembling the Social: An Introduction to Actor-Network Theory*. Oxford: Oxford University Press.

Latour, B. and Woolgar, S. 1979. *Laboratory Life: The Social Construction of Scientific Facts*. Beverly Hills, CA: Sage Publications.

Latour, B. and Woolgar, S. 1986. *Laboratory Life: The Construction of Scientific Facts*. Princeton, NJ: Princeton University Press.

Law, J. 1986. On the methods of long-distance control: vessels, navigation, and the Portuguese route to India, in *Power, Action and Belief. A New Sociology of Knowledge?*, edited by Law, J. London: Routledge, 234–263.

Law, J. 2003a. Ladbroke Grove, or How to Think About Failing Systems. Available at: http://www.lancs.ac.uk/fass/sociology/papers/law-ladbroke-grove-failing-systems.pdf. Accessed 20 February 2013.

Law, J. 2003b. Disasters, a/Symmetries and Interferences. Available at: http://www.lancs.ac.uk/fass/sociology/papers/law-disaster-asymmetries-and-interferences.pdf. Accessed 20 February 2013.

Letnes, B., Korsvold, T. and Larsen, S. 2008. *eDrilling – Expectations and Factors for Successful Implementation on Ekofisk*. Trondheim: Center for Integrated Operations in the Petroleum Industry.

Leveson, N. 2004. A new accident model for engineering safer systems. *Safety Science*, 42(4), 237–270.

Leveson, N., Dulac, N., Marais, K. and Carroll, J. 2009. Moving beyond normal accidents and high-reliability organizations: a systems approach to safety in complex systems. *Organization Studies*, 30(2–3), 227–249.

Lynch, M. 1982. Technical work and critical inquiry: investigations in a scientific laboratory. *Social Studies of Science*, 12(4), 499–533.

Løwén, S., Nygård, B.E., Østensen, S. and Lund, T. 2009. Subsurface Support Centre: A Hub for Communication of Knowledge. Paper to the SPE Digital Energy Conference and Exhibition, Houston, Texas, 7–8 April 2009.

MacMillan, J., Entin, E.E. and Serfaty, D. 2004. Communication overhead: the hidden cost of team cognition, in *Team Cognition: Understanding the Factors That Drive Process and Performance*, edited by Salas, E. and Fiore, S.M. Washington, DC: American Psychological Association, 61–82.

Marais, K., Dulac, N. and Leveson, N. 2004. Beyond Normal Accidents and High Reliability Organizations: The Need for an Alternative Approach to Safety in Complex Systems. Paper to the Engineering Systems Division Symposium, 29–31 March, MIT, Cambridge, MA.

Mintzberg, H. 1979. *The Structuring of Organizations: A Synthesis of the Research*. Englewood Cliffs, NJ: Prentice-Hall.

Moltu, B. and Nærheim, J. 2010. IO design gives high efficiency. *SPE Economics & Management*, 2(1), 32–37.

Munkvold, G. and Ellingsen, G. 2007. Common Information Spaces Along the Illness Trajectories of Chronic Patients. Paper to the 10th European Conference on Computer-Supported Cooperative Work, 24–28 September, Limerick, Ireland.

National Transportation Safety Board 1994. *Safety Study: A Review of Flightcrew-Involved Major Accidents of U.S. Air Carriers, 1978 through 1990*. Washington, DC: NTSB.

Nemeth, C., Hollnagel, E. and Dekker, S. 2009. *Resilience Engineering Perspectives: Preparation and Restoration*. Aldershot: Ashgate.

Nofi, A.A. 2000. *Defining and Measuring Shared Situational Awareness*. Alexandria, VA: Center for Naval Analyses.

Okhuysen, G. and Bechky, B. 2009. Coordination in organizations: an integrative perspective. *The Academy of Management Annals*, 3(1), 463–502.

OLF 2003. *eDrift på Norsk Sokkel – det Tredje Effektiviseringsspranget [eDrift on the Norwegian Continental Shelf – the Third Efficiency Leap]*. Stavanger: Oljeindustriens Landsforening.

OLF 2005. *Integrated Work Processes: Future Work Processes on the Norwegian Continental Shelf*. Stavanger: Oljeindustriens Landsforening.

Orr, J. 1996. *Talking About Machines: An Ethnography of a Modern Job*. New York: Cornell University Press.

Perrow, C. 1984. *Normal Accidents: Living with High-Risk Technologies*. Princeton, NJ: Princeton University Press.

Perrow, C. 1999. *Normal Accidents: Living with High-Risk Technologies*. Princeton, NJ: Princeton University Press.

Ptil 2007. *Human Factors i Bore- og Brønnoperasjoner. Borernes Arbeidssituasjon [Human Factors in Drilling and Well Operations. The Drillers' Working Situation]*. Stavanger: Petroleum Safety Authority Norway.

Randall, D. 2000. What's Common About Common Information Spaces. Paper to the Workshop on Cooperative Organisation of Common Information Spaces, August 2000, Technical University of Denmark.

Rasmussen, J. 1997. Risk management in a dynamic society: a modelling problem. *Safety Science*, 27(2–3), 183–213.

Reason, J. 1987. Cognitive aids in process environments: prostheses or tools? *International Journal of Man-Machine Studies*, 27(5–6), 463–470.

Reason, J. 1995. Understanding adverse events: human factors. *Quality in Health Care*, 4(2), 80–89.

Reason, J. 1997. *Managing the Risks of Organizational Accidents*. Aldershot: Ashgate.

Reason, J. 1998. Achieving a safe culture: theory and practice. *Work & Stress*, 12(3).

Reason, J. 2000. Human error: models and management. *BMJ*, 320(7237), 768–770.

Rijpma, J.A. 1997. Complexity, tight-coupling and reliability: connecting normal accidents theory and high-reliability theory. *Journal of Contingencies and Crisis Management*, 5(1), 15–23.

Rijpma, J.A. 2003. From deadlock to dead end: the normal accidents-high-reliability debate revisited. *Journal of Contingencies and Crisis Management*, 11(1), 37–45.

Ringstad, A.J. and Andersen, K. 2006. Integrated Operations and HSE – Major Issues and Strategies. Paper to the SPE International Conference on Health, Safety, and Environment in Oil and Gas Exploration and Production, Abu Dhabi, U.A.E., 2–4 April 2006.

Ringstad, A.J. and Andersen, K. 2007. Integrated Operations and the Need for a Balanced Development of People, Technology and Organisation. Paper to the International Petroleum Technology Conference, 4–6 December 2007, Dubai, U.A.E.

RIO project. 2010. Interdisciplinary Risk Assessment in Integrated Operations Addressing Human and Organisational Factors. Available at: http://www.sintef.no/Projectweb/RIO/Integrated-Operations.

Rochlin, G.I. 1999. Safe operation as a social construct. *Ergonomics*, 42(11), 1549–1560.

Rolland, K.H., Hepsø, V. and Monteiro, E. 2006. Conceptualizing Common Information Spaces across Heterogeneous Contexts: Mutable Mobiles and Side-effects of Integration, in Proceedings of the 2006 20th Anniversary Conference on Computer-Supported Cooperative Work. New York:ACM Press.

Rommetveit, R., Bjørkevoll, K.S., Halsey, G.W., Larsen, H.F., Merlo, A., Nossaman, L.N., Sweep, M.N., Silseth, K.M. and Ødegård, S.I. 2004. Drilltronics: An Integrated System for Real-Time Optimization of the Drilling Process. Paper to the IADC/SPE Drilling Conference, Dallas, Texas, 2–4 March 2004.

Rommetveit, R., Bjorkevoll, K.S., Odegaard, S., Herbert, M. and Halsey, G. 2008a. Automatic Real-Time Drilling Supervision, Simulation, 3D Visualization, and Diagnosis on Ekofisk. Paper to the SPE Conference, 4–6 March 2008, Orlando, Florida.

Rommetveit, R., Bjørkevoll, K.S., Ødegård, S.I., Herbert, M., Halsey, G.W. and Kluge, R. 2008b. eDrilling Used on Ekofisk for Real-Time Drilling Supervision, Simulation, 3D Visualization and Diagnosis. Paper to the SPE Intelligent Energy Conference and Exhibition, Amsterdam, 25–27 February 2008.

Rosendahl, T and Hepsø, V. 2013. *Integrated Operations in the Oil and Gas Industry: Sustainability and Capability Development*. Hershey, PA.: IGI Global.

Rosness, R., Grøtan, T.O., Guttormsen, G., Herrera, I.A., Steiro, T., Størseth, F., Tinmannsvik, R.K. and Wærø, I. 2010. *Organisational Accidents and Resilient Organisations: Six Perspectives*, Revision 2. Trondheim: SINTEF Technology and Society, Safety Research.

Roth, E.M., Multer, J. and Raslear, T. 2006. Shared situation awareness as a contributor to high reliability performance in railroad operations. *Organization Studies*, 27(7), 967–987.

Sagan, S.D. 1993. *The Limits of Safety: Organizations, Accidents, and Nuclear Weapons*. Princeton, NJ: Princeton University Press.

Schiefloe, P.M. 2009. Oljelandet, in *Det Norske Samfunn [The Norwegian Society]*, edited by Frønes, I. and Kjølsrød, L., Oslo, Norway: Gyldendal akademisk.

Schiefloe, P.M. and Vikland, K.M. 2009. Close to Catastrophe. Lessons from the Snorre A Gas Blow-Out. Paper to the 25th European Group for Organizational Studies (EGOS), Barcelona, Spain, 3–4 July 2010.

Schiefloe, P.M., Vikland, K.M., Ytredal, E.B., Torsteinsbø, A., Moldskred, I.O., Heggen, S., Sleire, D.H., Førsund, S.A. and Syversen, J.E. 2005. Årsaksanalyse Etter Snorre A-Hendelsen 28.11.*2004 [Causal Analysis of the Snorre A Incident 28.11.2004]*. Stavanger: Statoil.

Schmidt, K. 1996. Coordination mechanisms: towards a conceptual foundation of CSCW systems design. *Computer-Supported Cooperative Work*, 5(2), 155–200.

Schmidt, K. 2010. *Cooperative Work and Coordinative Practices: Contributions to the Conceptual Foundations of Computer Supported Cooperative Work (CSCW)*. New York: Springer-Verlag New York Inc.

Schmidt, K. and Bannon, L. 1992. Taking CSCW seriously: supporting articulation work. *Computer-Supported Cooperative Work*, 1(1), 7–40.

Schulman, P.R. 1993. The negotiated order of organizational reliability. *Administration Society*, 25(3), 353–372.

Shapin, S. and Schaffer, S. 1985. *Leviathan and the Air-Pump: Hobbes, Boyle, and the Experimental Life*. Princeton, NJ: Princeton University Press.

Shokouhi, S., Aamodt, A., Skalle, P. and Sørmo, F. 2009. Determining Root Causes of Drilling Problems by Combining Cases and General Knowledge. Paper to the Eighth International Conference on Case-Based Reasoning, ICCBR 2009, Seattle, Washington, 20–23 July, 2009.

Skarholt, K., Næsje, P., Hepsø, V. and Bye, A.S. 2008. Integrated operations and leadership – how virtual cooperation influences leadership practice, in *Safety, Reliability and Risk Analysis: Theory, Methods and Applications. Proceedings of the European Safety and Reliability Conference*, Esrel 2008, and Seventeenth SRA-Europe, edited by Martorell, S., Guedes Soares, C. and Barnett, J. London: Taylor & Francis Group, 821–828.

Snook, S.A. 2000. *Friendly Fire: The Accidental Shootdown of U.S. Black Hawks over Northern Iraq*. Princeton, NJ: Princeton University Press.

Stanton, N., Stewart, R., Harris, D., Houghton, R., Baber, C., McMaster, R., Salmon, P., Hoyle, G., Walker, G. and Young, M. 2006. Distributed situation awareness in dynamic systems: theoretical development and application of an ergonomics methodology. *Ergonomics*, 49(12), 1288–1311.

Star, S.L. and Griesemer, J.R. 1989. Institutional ecology, 'translations' and boundary objects: amateurs and professionals in Berkeley's Museum of Vertebrate Zoology, 1907–39. *Social Studies of Science*, 19(3), 387–420.

Star, S.L. and Strauss, A. 1999. Layers of silence, arenas of voice: the ecology of visible and invisible work. *Computer-Supported Cooperative Work*, 8(1–2), 9–30.

Stortinget 2004. *Stortingsmelding Nr. 38. Om Petroleumsvirksomheten [On the Petroleum Activity]*. Oslo: The Norwegian Parliament.

Strauss, A. 1985. Work and the division of labor. *Sociological Quarterly*, 26(1), 1–19.

Strum, S.S. and Latour, B. 1987. Redefining the social link: from baboons to humans. *Social Science Information*, 26(4), 783–802.

Suchman, L. 1987. *Plans and Situated Actions: The Problem of Human–Machine Communication*. Cambridge: Cambridge University Press.

Suchman, L. 1993. Centers of coordination: a case and some themes, in *Discourse, Tools, and Reasoning: Essays on Situated Cognition*, edited by Resnick, L.B. Berlin: Springer, 41–62.

Suchman, L. 1995. Making work visible. *Communications of the ACM*, 38(9), 56–64.

Suchman, L. 1996a. Constituting shared workspaces, in *Cognition and Communication at Work*, edited by Engeström, Y. and Middleton, D. Cambridge: Cambridge University Press, 35–60.

Suchman, L. 1996b. Supporting articulation work, in *Computerization and Controversy: Value Conflicts and Social Choices*, edited by Kling, R. San Diego, CA: Academic Press, 407–423.

Suchman, L. 2007. *Human-Machine Reconfigurations: Plans and Situated Actions*. New York: Cambridge University Press.

Suchman, L.A. 1988. Representing practice in cognitive science. *Human Studies*, 11(2), 305–325.

Suwartadi, E., Krogstad, S. and Foss, B. 2010. A Lagrangian-Barrier Function for Adjoint State Constraints Optimization of Oil Reservoirs Water Flooding, in Proceeding of IEEE Conference on Decision and Control, 15–17 December 2010, Atlanta, Georgia, USA.

Tinmannsvik, R.K. 2008. *Robust Arbeidspraksis: Hvorfor Skjer det Ikke Flere Ulykker på Sokkelen? [Resilient Work Practices: Why Do Not More Accidents Occur on the Norwegian Continental Shelf?]* Trondheim: Tapir akademisk forlag.

Tjora, A.H. 2000. The technological mediation of the nursing–medical boundary. *Sociology of Health & Illness*, 22(6), 721–741.

TRADOC. 1995. *Tradoc Pamphlet 525–69. Military Operations: Concept for Information Operations*. Available at: http://www.iwar.org.uk/iwar/resources/tradoc/p525-69.htm.

Traweek, S. 1992. *Beamtimes and Lifetimes: The World of High-Energy Physicists*. Cambridge, MA: Harvard University Press.

Turner, B.A. 1976. The organizational and interorganizational development of disasters. *Administrative Science Quarterly*, 21(3), 378–397.

Turner, B.A. 1978. *Man-Made Disasters*. London: Wykeham.

Vaidya, P. and Rausand, M. 2010. Technical Health of a System – in the Context of Condition Based Maintenance. Paper to The European Safety and Reliability Association Annual Conference (ESREL), 5–9 September 2010, Rhodos, Greece.

Vaughan, D. 1996. *The Challenger Launch Decision: Risky Technology, Culture, and Deviance at Nasa*. Chicago, IL: University of Chicago Press.

Veland, Ø. and Andresen, G. 2011. Demonstration of a Research Prototype of a Collaborative Planning Tool for Use in Offshore Petroleum Operations. Paper to the European Conference on Computer-Supported Cooperative Work (ECSCW), 24–28 September, 2011, Aarhus, Denmark.

Vinge, H. 2009. *Diskurser om Integrerte Operasjoner: En Sosiologisk Studie av et Endringskonsept i Petroleumsbransjen [Discourses on Integrated Operations: A Sociological Study of a Change Concept in the Petroleum Industry]*. Trondheim: Norwegian University of Science and Technology.

Weick, K. 1995. What theory is not, theorizing is. *Administrative Science Quarterly*, 40(3), 385–390.

Weick, K.E. 1987. Organizational culture as a source of high reliability. *California Management Review*, 29(2), 112–128.

Weick, K.E. 2007. The generative properties of richness. *Academy of Management Journal*, 50(1), 14.

Weick, K.E. and Roberts, K.H. 1993. Collective mind in organizations: heedful interrelating on flight decks. *Administrative Science Quarterly*, 38(3), 357–381.

Weick, K.E. and Sutcliffe, K.M. 2001. *Managing the Unexpected: Assuring High Performance in an Age of Complexity*. San Francisco, CA: Jossey-Bass.

Weick, K.E. and Sutcliffe, K.M. 2007. *Managing the Unexpected: Resilient Performance in an Age of Uncertainty*. San Francisco, CA: Jossey-Bass.

Weick, K.E. and Sutcliffe, K.M. 2008. Information overload revisited, in *The Oxford Handbook of Organizational Decision Making*, edited by Hodgkinson, G.P. and Starbuck, W.H. Oxford: Oxford University Press, 56–75.

Weick, K.E., Sutcliffe, K.M. and Obstfeld, D. 1999. Organizing for high reliability: processes of collective mindfulness. *Research in Organizational Behavior*, 21, 81–123.

Woods, D.D. 2006. Essential characteristics of resilience, in *Resilience Engineering: Concepts and Precepts*, edited by Hollnagel, E., Woods, D.D. and Leveson, N. Aldershot: Ashgate, 21–34.

Index